AF651260

BASICS OF PHYSICAL CHEMISTRY

BASICS OF
PHYSICAL CHEMISTRY

Edited by
Prabir Pujhari

ANMOL PUBLICATIONS PVT. LTD.
NEW DELHI - 110 002 (INDIA)

ANMOL PUBLICATIONS PVT. LTD.

H.O.: 4374/4B, Ansari Road, Darya Ganj,
New Delhi-110 002 (India)
Ph.: 23278000, 23261597

B.O.: No. 1015, Ist Main Road, BSK IIIrd Stage
IIIrd Phase, IIIrd Block,
Bangalore - 560 085 (India)
Visit us at: www.anmolpublications.com

Basics of Physical Chemistry

© Author

First Published, 2008

ISBN 978-81-261-3548-6

PRINTED IN INDIA

Printed at Mehra Offset Press, Delhi.

Contents

Preface

The purpose of the present work is primarily that of considering the development of certain branches of chemical science. The method of treatment, namely that of selecting particular topics and allocating them to separate chapters—that is, division according to subject-matter and not according to periods of time—has again been adopted. It may be added—as is indicated by the title—that somewhat greater emphasis has been given to certain of the older parts of the science.

Although the book is intended for the serious student of chemistry, the style of treatment is 'elementary' in the sense that it should appeal to others whose acquaintance with the subject is more limited, and whose chief interests may lie in altogether different kinds of study. It will doubtless be appreciated that in a book of this kind the sketches of individual topics must of necessity be somewhat impressionistic.

Author

Chapter 1

Statistical Thermodynamics

Physical chemistry is the application of the principles and methods of physics and math to chemistry. Physical chemistry can also be regarded as the study of the physical principles underlying chemistry. We want to know how and why materials behave as they do.

The ultimate goal of physical chemistry is to provide a (mathematical) model for all of chemistry.

Physical chemistry requires that calculus be used as a tool just as algebra has been used as a tool in previous courses.

The derivations and calculations of physical chemistry require lots of partial derivatives. (This is because the functions we deal with are functions of several variables.) We will also do lots of simple integrals. In the first semester the integrals are mostly in one variable. In the second semester there will be more integrals in two and three dimensions.

Thermodynamics is what we call a macroscopic theory. That is, it deals with the bulk properties of matter and does not concern itself with whether or not there are atoms or molecules. In fact, thermodynamics does not care whether or not there are atoms and molecules. On the other hand, quantum mechanics is a microscopic theory because it deals with the individual particles of matter. Statistical thermodynamics brings us full circle by providing a mechanism for calculating the properties of bulk material (macroscopic samples) from the properties of the atoms and molecules which comprise the material.

(Recently there has been a lot of interest in mesoscopic materials. These are materials which are composed of relatively small numbers of particles. They consist of so few particles that they do not manifest the same properties as the bulk matter, yet they have enough particles that they no longer have the properties of individual atoms or molecules. Work in this area has given rise to the so-called "nanoscale" technologies.)

MATTER - STATES OF MATTER

Matter is anything that has mass and takes up space (to use an old freshman chemistry definition). In physical chemistry we will mainly be concerned with matter that is built up from protons, neutrons and electrons. We rarely will be concerned with the more exotic forms of matter like positrons and mesons and essentially never with quarks. The matter we are most concerned with is made when protons, neutrons and electrons are put together to form atoms and molecules.

Matter exists in several possible states. The most common states of matter on the surface of our planet are: solid; liquid; gas.

However, there are other states of matter: plasma (ionized gases); nuclear matter (as in neutron stars); white dwarf stars; interfacial matter (Material at surfaces often has different properties than bulk matter.);"black hole" matter etc.

Most of the matter in the universe is not in one of the states, solid, liquid, or gas, but in one of the more exotic states like plasma. Even in our solar system solid, liquid and gas are the minority forms of matter. The solar system is dominated by the sun and the sun is mostly a plasma. (Molecular water has been detected in sun spots, which are relatively cool portions of the sun's "surface.")

VARIABLES TO DESCRIBE MATTER

We can describe a sample of matter by using variables such as mass, number of moles, volume, temperature, pressure, density and so on. we usually symbolize these variables as (respectively) *m, n, V, T, p, ?* (lower case Greek "rho"), and so on. These variables are called "state variables" because they describe the state of the system and because they depend only on the state of the system. We will be defining more state variables as we go along.

Variables describing matter can be divided into two classes. Variables whose value is proportional to the amount of sample are called extensive variables and variables which are independent of the amount of sample are called intensive variables. (You can remember these by letting the word extensive remind you of the word "extent and letting intensive remind you of intensity and vice versa.)

In the above list you should convince yourself that *m, n,* and *V* are extensive and *T, p,* and ? are intensive.

The state of a system is given by specifying the values of all the variables describing the system. This definition of "state" assumes that the system is at equilibrium. We will give a proper thermodynamic definition later, but for now we define equilibrium by saying that everything that wants to happen in the system has happened. Another way to say this is to say that the system has the properties it would have after infinite time. You could say that equilibrium is the state when none of the values of the variables is changing in time, but here you have to be careful to exclude "steady-state" systems. (An example of a steady-state system is one where material is flowing in and out of the system, but the system itself appears not to be changing.)

UNITS AND DIMENSIONS

We will mostly use the SI (Système International d'Unités) system of units. Some exceptions are listed below. The SI system is the standard system of units in the world today. It is an outgrowth of the old mks system. The system is metric in nature, meaning that larger or smaller units are obtained by multiplying or dividing a base unit by powers of ten. There is an excellent description of the SI system, with all of the details (including prefixes), at the NIST website. (NIST also has a list of all of the fundamental constants.)

We will use some non-SI units. It is important to know that all non-SI units are now defined in terms of SI units.

For volume we will use liters (L) and mL. There are 1000 L in a m^3.

For energy we will occasionally see liter atmospheres (Latm) or liter bars (Lbar). You need to convince yourself that a pressure times a volume has units of energy. Some of our calculations will give us answers in Latm. These should always be converted to Joules.

1 Latm = 101.325 J

1 Lbar = 100 J

Occasionally we see the energy unit calorie (cal), not to be confused with the dietary Calorie (Cal) which is really a kcal. The calorie is defined as 1 cal = 4.184 J exactly.

The SI unit of pressure is the Paschal (Pa). The Pa is a force of 1 Newton per m^2. Other pressure units are atmospheres, bars and Torr. The conversions between these units are

1 atm = 101325 Pa = 1.01325 bar = 760 Torr.

The Torr is written in the older literature as mmHg.

We will not often use English units for length, but it is sometimes useful to know that the inch is defined as exactly 0.0254 m (.or 2.54 cm).

THE STATE OF A SYSTEM

We specify the state of a system - say, a sample of material - by specifying the values of all the variables describing the system. If the system is a sample of a pure substance this would mean specifying the values of the temperature, T, the pressure, p, the volume, V, and the number of moles of the substance, n.

(We must assume that the system is at equilibrium. That is, none of the variables is changing in time and they have the values they would have if we let time go to infinity. We will give a thermodynamic definition of equilibrium later, but this one will suffice for now.)

(If the system is a mixture you also have to specify the composition of the mixture as well as T, p, and V. This could be done by specifying the number of moles of each component, $n_1, n_2, n_3, \ldots$, or by specifying the total number of moles of all the substances in the mixture and the mole fraction of each component, $X_1, X_2, X_3, \ldots$. We will not deal with mixtures on this page.)

EQUATIONS OF STATE

Let's consider a sample of a pure substance, say n moles of the substance. It is an experimental fact that the variables, T, p, V, and n are not independent of each other. That is, if we change one variable one (or more) of the other variables will change too. This means that there must be an equation connecting the variables. In other words, there is an equation that relates the variables to each other. This equation is called the "equation of state." The most general form for an equation of state is,

$$f(p,V,T,n) = 0 \tag{1}$$

This equation is not very useful because it does not tell us the detailed form of the function, f. However, it does tell us that we should be able to solve the equation of state for any one of the variables in terms of the other three. For example, we can, in principle, find

$$p = p(V,T,n) \tag{2}$$

or

$$V = V(p,T,n) \tag{3}$$

and so on. (These last two equations should be read as, "p is a function of V, T, and n" and "V is a function of p, T, and n.")

If we do some more experiments we will notice that when we hold p and T constant we can't change n without changing V and vice versa. In fact, V is proportional to n. That is, if we double n, the volume, V, will also double, and so on. Because V is proportional to n these two variables must always appear in the equation of state as V/n (or n/V). This means that the most general form of the equation of state is simpler than that shown above. The most general form of the equation of state really has the form,

$$f\left(p,\frac{V}{n},T\right)=0 \tag{4}$$

which can be solved for p, V/n, or T in terms of the other two. For example,

$$p = p\left(\frac{V}{n},T\right) \tag{5}$$

and so on.

All isotropic substances have, in principle, an equation of state, but we do not know the equation of state for any real substance. All we have is some approximate equations of state which are useful over a limited range of temperatures and pressures. Some of the approximate equations of state are pretty good and some are not so good. Our best equations of state are for gases. There are no general equations of state for liquids and solids, isotropic or otherwise. On another page we will show you how to obtain an approximate equation of state for isotropic liquids and solids which is acceptable for a limited range of temperatures around 25°C and for a limited range of pressures near one atmosphere.

THE IDEAL GAS EQUATION OF STATE

The best known equation of state for a gas is the "ideal gas equation of state." It is usually written in the form,

$$pV = nRT \tag{6}$$

This equation contains a constant, R, called the gas constant or, sometimes, the universal gas constant[2]. We can write this equation in the forms shown above if we wish. For example, the analog of Equation (1) is,

$$pV = nRT \tag{7}$$

The analogs of Equations (2) and (3) are.

$$p = \frac{nRT}{V} \tag{8}$$

and

$$V = \frac{nRT}{p} \tag{9}$$

respectively, and so on.

No real gas obeys the ideal gas equation of state for all temperatures and pressures. However, all gases obey the ideal gas equation of state in the limit as pressure goes to zero (except possibly at very low temperatures). Another way to say this is to say that all gases become ideal in the limit of zero pressure.

The ideal gas equation of state is the consequence of a model in which the molecules are point masses - that is, they have no size - and in which there are no attractive forces between the molecules.

THE VAN DER WAALS EQUATION OF STATE

The van der Waals equation of state is,

$$(p + a\frac{n^2}{V^2})(V - nb) = \text{nRT} \tag{10}$$

Notice that the van der Waals equation of state differs from the ideal gas by the addition of two adjustable parameters, a, and b (among other things). These parameters are intended to correct for the omission of molecular size and intermolecular attractive forces in the ideal gas equation of state. The parameter b corrects for the finite size of the molecules and the parameter, a, corrects for the attractive forces between the molecules.

The argument goes something like this: Assume that an Avogadro's number of molecules (i.e., a mole of the molecules) takes up a volume of space - just by their physical size - of b Liters. Then any individual molecule doesn't have the whole (measured) volume, V, available to move around in. The space available to any one molecule is just the measured volume less the volume taken up by the molecules themselves, nb. So the "effective" volume, which we shall call V_{eff}, is $V - nb$. The effective pressure, p_{eff}, is a little bit trickier. Consider a gas where the molecules attract each other. The molecules at the edge of the gas (near the container wall) are attracted to the interior molecules. The number of "edge" molecules is proportional to n/V and the number of interior molecules is proportional to n/V also. The number of pairs of interacting molecules is thus proportional to n^2/V^2 so that the forces attracting the edge molecules to the interior are proportional to n^2/V^2. These forces give an additional contribution to the pressure on the gas proportional to n^2/V^2. We will call the proportionality constant a so that the effective pressure becomes,

$$peff = p + a\frac{n^2}{V^2}$$

We now guess that the gas would obey the ideal gas equation of state if only we used the effective volume and pressure instead of the measured volume and pressure. That is,

$$p_{eff}V_{eff} = nRT \quad (11)$$

Inserting our forms for the effective pressure and volume we get,

$$(p + a\frac{n^2}{V^2})(V - nb) = \text{nRT} \tag{12}$$

which is the van der Waals equation of state.

The van der Waals constants, *a* and *b*, for various gases must be obtained from experiment or from some more detailed theory. They are tabulated in handbooks and in most physical chemistry textbooks.

1. Isotropic means that the properties of the material are independent of direction within the material. All gases and most liquids are isotropic, but crystals are not. The properties of the crystal may depend on which direction you are looking with respect to the crystal lattice. As we said above, most liquids are isotropic, but liquid crystals are not. That's why they are called liquid crystals. Amorphous solids and polycrystalline solids are usually isotropic.
2. The value of the gas constant, *R*, depends on the units being used.

   R = 8.314472 J/Kmol

   = 0.08205746L atm/Kmol

   = 1.987207 cal/K mol

   = 0.08314472 L bar/Kmol.

THE MAXWELL-BOLTZMANN DISTRIBUTION

The molecular description of the bulk properties of a gas depends upon our knowing the mathematical form of the velocity distribution; That is, the probability, $F(V_x, V_y, V_z)\Delta V_x$, ΔV_y, ΔV_z, of finding a molecule with velocity components in the range v_x to $v_x + \Delta vx$, v_y to $v_y + \Delta v_y$, and v_z to $v_z + \Delta v_z$. This can be found by making two very simple assumptions:

1. The three components of the velocity are independent of each other. This implies that the velocity probability density can be written as:

$$F(v_x, v_y, v_x) = f(v_x)f(v_x)f(v_y)f(v_z) \tag{1}$$

2. All directions are equivalent (space is isotropic). This implies that the probability of finding a molecule moving with a certain velocity cannot depend on the direction; it is equally probable to find a molecule with any speed v moving in any direction. Mathematically, this implies that the probability density can only depend on the magnitude of the velocity, or the molecular speed. Hence,

$$F(v_x, v_y, v_x) = f([v_x^2 + v_y^2 + v_z^2]^{1/2}) \tag{2}$$

By comparing Eqs. (1) and (2), we have

$$f(v_x)F(v_y)F(v_z) = f([v_x^2 + v_y^2 + v_z^2]^{1/2}) \tag{3}$$

Very few functions can satisfy Eq. (3); in fact, the only one is:

$$f(v_i) = Ae^{-bv_i^2},\ i = x, y, z \tag{4}$$

where A and b are, as yet, arbitrary constants. Verify that f given by Eq. (4) satisfies Eq. (3). To show that Eq. (4) is the only possible function, requires some mathematics which might be beyond the level of Chemistry 213; however, the proof is contained in the appendix for those of you who have this extra background.

How do we determine the values of the constants A and b? The function f, is a probability density and every molecule must have some velocity. From the previous section, we know that this means

$$1 = \sum_{V_x=-\infty}^{\infty} f(v_x)\Delta v_x \rightarrow \int_{-\infty}^{\infty} f(v_x)dv_x, \tag{5}$$

where the integral is obtained when we make Dv_x very small. By going to a table of integrals we find that

$$\int_{-\infty}^{\infty} Ae^{-bv^2}dv = A\left[\frac{\pi}{b}\right]^{1/2}, \tag{6}$$

which when used in Eq. (5) gives

$$A = \left[\frac{b}{\pi}\right]^{1/2}, \tag{7}$$

The parameter b is found as follows: We will compute the pressure that a dilute gas exerts on the walls of its container and then compare the result with experiment (i.e., the ideal gas equation of state). What is the pressure and how do we get it from the velocity distribution? The pressure is the force exerted by the gas per unit area. Moreover, the force is the rate of change of momentum per unit time due to collisions with the wall.

Imagine the molecules which comprise our gas as billiard balls and assume that the walls are perfectly smooth. What happens when a molecule collides with the wall? Actually relatively little; the normal component of the velocity changes sign .

If the wall is taken to be the y-z plane, the momentum change is

$$\Delta p = -2mv_x, \tag{8}$$

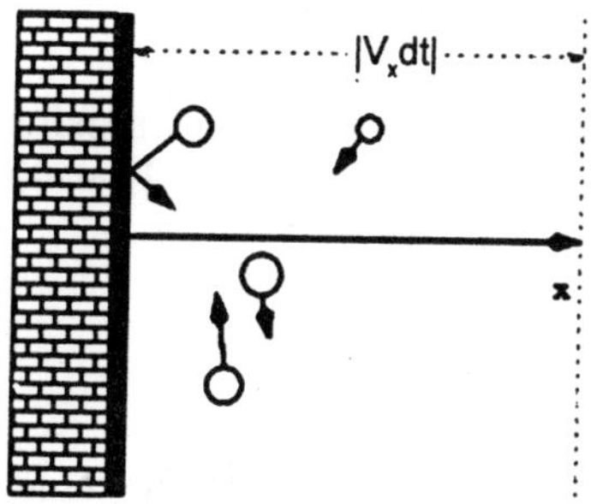

Fig.

where m is the mass of the molecule. A typical velocity is 10^5 cm/sec; what is a typical momentum change for argon? How many v_x molecules with velocity will collide with the wall per unit area in time Δt? From fig., it should be clear that any molecule within a distance$v_x\Delta t$ will hit the wall. The number per unit area is therefore:

$$n(v_x)v_x\Delta t \tag{9}$$

Where $n(v)$ is the number of molecules per unit volume with x component of velocity in the range v_x to $v_x + \Delta v_x$. This is related to the molecular velocity distribution by

$$n(v_x) = n_0 f(v_x)\Delta v_x, \tag{10}$$

Where n_0 is the number of molecules per unit volume. By multiplying Eqs. (8) and (9), dividing by Δt and by using Eq. (10), we arrive at the pressure exerted by those molecules with x velocity component in the range v_x to $v_x + \Delta v_x$:

$$2mn_0 v_x^2 f(v_x)\Delta v_x \tag{11}$$

All that remains is to include the contributions from all velocities corresponding to molecules moving towards the wall. The pressure, P, thus equals:

$$p = \sum_{v_x \geq 0}^{\infty} 2mn_0 v_x 2 f(v_x) dv_x \sum_{v_x \geq 0}^{\infty} = \int_0^{\infty} 2mn_0 v_x^2 f(v_x) dv_x$$

$$= \int_0^{\infty} 2mn_0 v_x^2 \left[\frac{b}{x}\right]^{1/2} e^{-bv_x^2} dv_x, \tag{12}$$

Where the last equality comes from using the explicit form of the probability density, [see Eqs. (4) and (7)]. The value of the integral is:

$$\int_0^{\infty} v_x^2 e^{-bv_x^2} dv = \frac{1}{4b}\left[\frac{\pi}{b}\right]^{1/2}.$$

If we use this result in Eq. (12), we find that

$$P = \frac{mn_0}{2b} \tag{13}$$

Next we write the ideal gas equation of state in terms of the number density:

$$P = n_0 k_B T,$$

Where k_B is Boltzmann's constant:

$$k_B = \frac{R}{6.0225 \times 10^{23}} = 1.38 \times 10^{-16}\ erg/deg.$$

By comparing this with Eq. (13) we see that

$$b = \frac{m}{2k_B T}.$$

The velocity probability density for the x component can now be written as

$$f(vx) = \left[\frac{m}{2\pi k_B T}\right]^{1/2} e^{-\frac{mv_x^2}{2k_B t}} \tag{14}$$

or for the full velocity as

$$F\left(v_x, v_y, v_x\right) = \left[\frac{m}{2\pi k_B T}\right]^{3/2} e^{-\frac{E}{k_B T}}, \tag{15}$$

where $E = m\left(v_x^2 + v_y^2 + v_z^2\right)/2$ is the energy of the molecule. This is referred to as the Maxwell-Boltzmann distribution function and is illustrated in the following figure:

Velocity Distribution

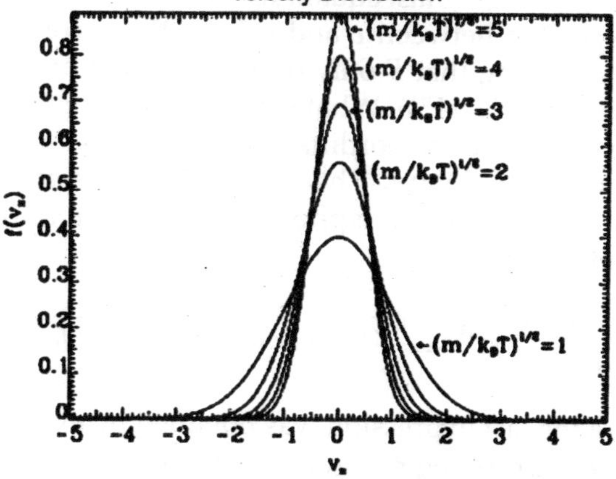

Notice that increasing the temperature makes it more probable to find molecules with higher velocities and that the probability of observing any given velocity will decrease as the energy associated with that velocity increases. This last comment is not limited to our simple billiard ball model of the molecules.

For many applications, the Maxwell-Boltzmann velocity distribution gives too much detail. In particular, remember that it is a probability density for the vector velocity. Suppose we are interested in some property which depends only on the speed of the molecules and not their direction. What is the probability density which describes the distribution of molecular speeds?

The speed distribution function, $F(c)dc$, is the probability of observing a molecule with a speed in the interval c to $c + dc$ irrespective of its direction. It can be obtained from Eq. (15) as follows: Eq. (15) describes the distribution of velocities. If we are not interested in the direction of the velocity, all we need do is to average it out; i.e.,

$$F(c)dc = \sum_{c \le |\vec{v}| \le c+dc} F\left(v_x, v_y, v_z\right) \Delta v_x \Delta v_y \Delta v_z$$

$$= F\left(v_x, v_y, v_z\right)\Big|_{|v|=c} \sum_{c \le |\vec{v}| \le c+dc} \Delta v_x \Delta v_y \Delta v_z,$$

The sums in these last equations are over all velocities such that the speed is between c and $c + dc$. The second equality comes from noting that F only depends on the magnitude of the velocity (see Eq. (15)). What is this last sum? $\Delta v_x\, \Delta v_y\, \Delta v_z$ is a volume element in a coordinate system whose axes are the components of velocity. The sum represents the volume between two concentric spheres of radius c and $c + dc$, respectively. Thus

$$\sum_{c \le |\vec{v}| \le c+dc} \Delta v_x \Delta v_y \Delta v_z = \frac{4\pi}{3}\left[(c+dc)^3 - c^3\right] \approx 4\pi c^2 dc, \tag{17}$$

where the last expression was obtained by expanding the products and dropping nonlinear terms in dc (remember that we will consider infinitesimally small dc). If we use Eq. (17) in (16), we find that

$$F(c) = 4\pi c^2 F\left(v_x, v_y, v_z\right)\Big|_{|v|=c} = 4\pi c^2 \left[\frac{m}{2\pi k_B T}\right]^{3/2} e^{-\frac{mc^2}{2k_g T}} \qquad (18)$$

This is the speed distribution and is shown below. Notice that it vanishes when c=0; why?

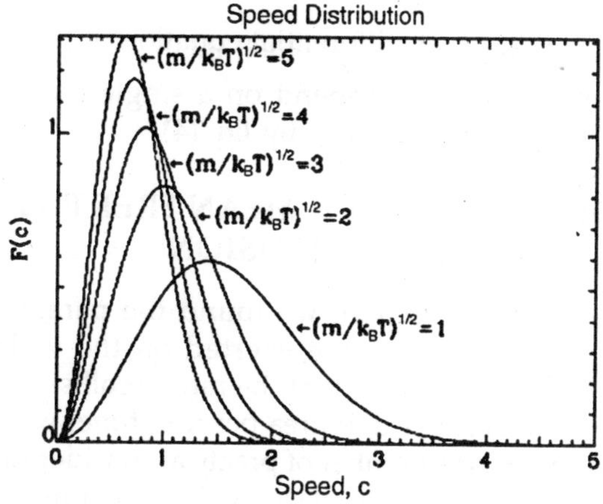

Appendix: Proof of Eq. (4)

You will not be responsible for this proof. Differentiate both sides of Eq. (3) with respect to and divide by v_x. You get:

$$\frac{\dfrac{df(v_x)}{dv_x} f(v_y) f(v_z)}{v_x} = \frac{\dfrac{dF\left(\left[v_x^2 + v_y^2 + v_z^2\right]^{1/2}\right)}{dv_x}}{\left[v_x^2 + v_y^2 + v_z^2\right]^{1/2}}.$$

By repeating this for v_y, equating the results and carrying out a little algebra, we find that:

$$\frac{\dfrac{df(v_x)}{dv_y}}{v_x f(v_x)} = \frac{\dfrac{df(v_y)}{dv_y}}{v_x f(v_y)}$$

The right hand side of the equation depends only on v_y and the left hand side depends only on v_x. Since these two components of velocity can be chosen independently, the only way that this last equation can be satisfied is if both sides are equal to the same constant. Hence,

$$\frac{dy(v_x)}{dv_x} = -\frac{1}{2}bv_x f(v_x).$$

This is a first order differential equation. It has a general solution which will depend on a single multiplicative constant. This solution is given by Eq. (4).

COLLISION RATES, MEAN FREE PATH AND DIFFUSION

In the previous section, we found the parameter b by computing the average force exerted on the walls of the container. Suppose, instead, that the rate of collisions (i.e., the number of collisions per unit area per unit time) was desired. This is important for a number of practical considerations; e.g., if a chemical reaction takes place every time a molecule hits the surface, then the rate of the reaction will just be the collision rate.

We obtain the collision rate by repeating the analysis which determined the force on the wall in the previous section. Thus the number of molecules with velocity v_x which collide in time interval $\Delta t$is

$$n(vx)vx\Delta t, \tag{1}$$

where we are using the same notation as in the preceding section. The total number of collisions thus becomes:

$$\begin{aligned} z_{wall}\Delta t &= \int_{-\infty}^{0} n(v_x)|v_x|\Delta t dv_x, \\ &= \int_{-\infty}^{0} dv_x n_0 \left(\frac{m}{2\pi k_B T}\right)^{1/2} . \, e^{-mv_x^2/(2kgT)|v_x|\Delta t} \\ &= \left(\frac{k_B T}{2\pi m}\right)^{1/2} n_0 \Delta t \end{aligned} \tag{2}$$

Hence the wall collision rate, z_{wall}, is given by

$$z_{wall} = \left(\frac{k_B T}{2\pi m}\right)^{1/2} n_0 \tag{3}$$

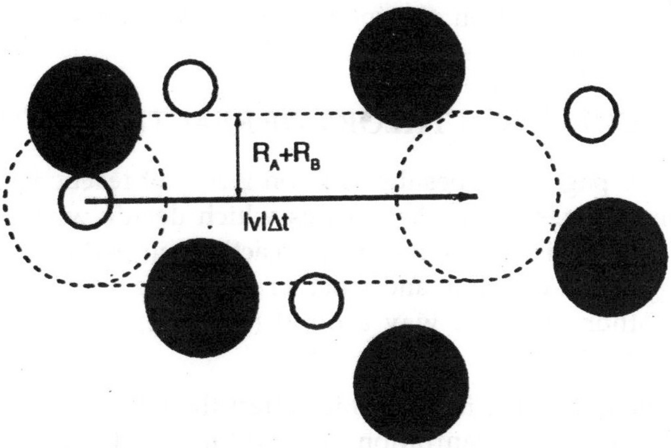

Fig.

Aside from the previously mentioned example of chemical reaction on a surface, another application of this expression is in *effusion* though a pinhole. If there is a hole of area A in the surface, the rate that molecules escape the system is just $Z_{wall}A$.

Notice that heavier molecules or isotopes will escape more slowly (with a $1/\sqrt{mass}$ dependence); hence, effusion through a pinhole is a simple way in which to separate different mass species.

Next consider the number of collisions which a molecule of type A makes with those of type B in a gas. We will model the two molecules as hard spheres of radii R_A and R_B, respectively. Moreover, in order to simplify the calculation, we will assume that the B molecules are stationary (this is a good approximation if $mB \gg mA$). An A molecule moving with velocity $\bar{v}$ for time Δtwill collide with any B molecule in a cylinder of radius $R_A + R_B$ and length $|\bar{v}|\Delta t$. Hence, volume swept out will just

$$|\bar{v}|\Delta t\pi\left(R_A + R_B\right)^2, \tag{4}$$

be and the actual number of collisions will $n_B \times$ volume swept out The average A–B collision rate, is obtained by multiplying by the probability density for A molecules with velocity $\bar{v}$ and averaging. Moreover, since only the speed of the molecule is important, we may use the molecular speed distribution function obtained in the preceding section to average over the speed of A.

THE COLLISION THEORY OF REACTION RATES

This page describes the collision theory of reaction rates. It concentrates on the key things which decide whether a particular collision will result in a reaction - in particular, the energy of the collision, and whether or not the molecules hit each other the right way around (the orientation of the collision).

The individual factors which affect the rate of a reaction (temperature, concentration, and so on) are discussed on separate pages. You can get at these via the rates of reaction menu - there is a link at the bottom of the page.

We are going to look in detail at reactions which involve a collision between two species.

Reactions where a single species falls apart in some way are slightly simpler because you won't be involved in worrying about the orientation of collisions. Reactions involving collisions between more than two species are going to be extremely uncommon.

REACTIONS INVOLVING COLLISIONS BETWEEN TWO SPECIES

It is pretty obvious that if you have a situation involving two species they can only react together if they come into contact with each other. They first have to collide, and then they *may* react.

Why "*may* react"? It isn't enough for the two species to collide - they have to collide the right way around, and they have to collide with enough energy for bonds to break.

(The chances of all this happening if your reaction needed a collision involving more than 2 particles are remote. All three (or more) particles would have to arrive at exactly the same point in space at the same time, with everything lined up exactly right, and having enough energy to react. That's not likely to happen very often!)

THE ORIENTATION OF COLLISION

Consider a simple reaction involving a collision between two molecules - ethene, $CH_2{=}CH_2$, and hydrogen chloride, HCl, for example. These react to give chloroethane.

$$CH_2{=}CH_2 + HCl \longrightarrow CH_3CH_2Cl$$

As a result of the collision between the two molecules, the double bond between the two carbons is converted into a single bond. A hydrogen atom gets attached to one of the carbons and a chlorine atom to the other.

The reaction can only happen if the hydrogen end of the H-Cl bond approaches the carbon-carbon double bond. Any other collision between the two molecules doesn't work. The two simply bounce off each other.

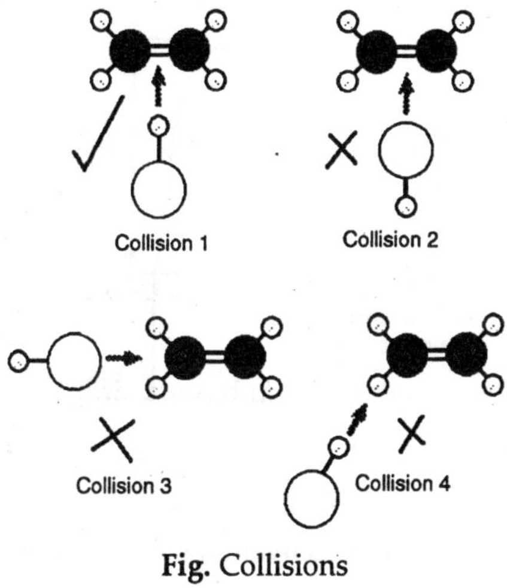

Fig. Collisions

Of the collisions shown in the diagram, only collision 1 may possibly lead on to a reaction.

If you haven't read the page about the mechanism of the reaction, you may wonder why collision 2 won't work as well. The double bond has a high concentration of negative charge around it due to the electrons in the bonds. The approaching chlorine atom is also slightly negative because it is more electronegative than hydrogen. The repulsion simply causes the molecules to bounce off each other.

In any collision involving unsymmetrical species, you would expect that the way they hit each other will be important in deciding whether or not a reaction happens.

THE ENERGY OF THE COLLISION

Activation Energy

Even if the species are orientated properly, you still won't get a reaction unless the particles collide with a certain minimum energy called the *activation energy* of the reaction.

Activation energy is the minimum energy required before a reaction can occur. You can show this on an *energy profile* for the reaction. For a simple over-all exothermic reaction, the energy profile looks like this:

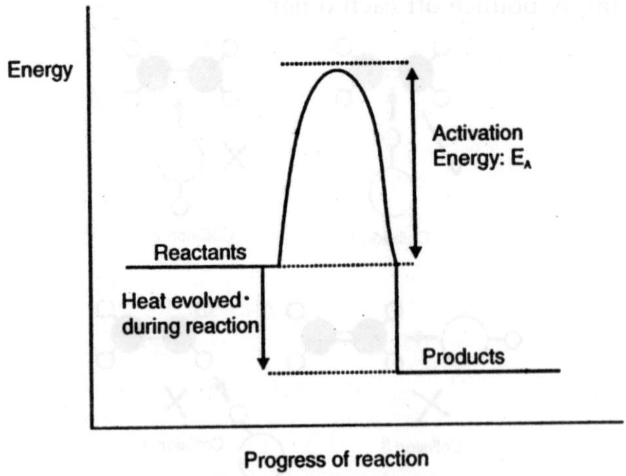

Fig. Activation Energy

If the particles collide with less energy than the activation energy, nothing important happens. They bounce apart. You can think of the activation energy as a barrier to the reaction.

Only those collisions which have energies equal to or greater than the activation energy result in a reaction.

Any chemical reaction results in the breaking of some bonds (needing energy) and the making of new ones (releasing energy). Obviously some bonds have to be broken before new ones can be made. Activation energy is involved in breaking some of the original bonds.

Where collisions are relatively gentle, there isn't enough energy available to start the bond-breaking process, and so the particles don't react.

THE MAXWELL-BOLTZMANN DISTRIBUTION

Because of the key role of activation energy in deciding whether a collision will result in a reaction, it would obviously be useful to know what sort of proportion of the particles present have high enough energies to react when they collide.

In any system, the particles present will have a very wide range of energies. For *gases*, this can be shown on a graph called the Maxwell-Boltzmann Distribution which is a plot of the number of particles having each particular energy.

Note: The graph only applies to *gases*, but the conclusions that we can draw from it can also be applied to reactions involving liquids.

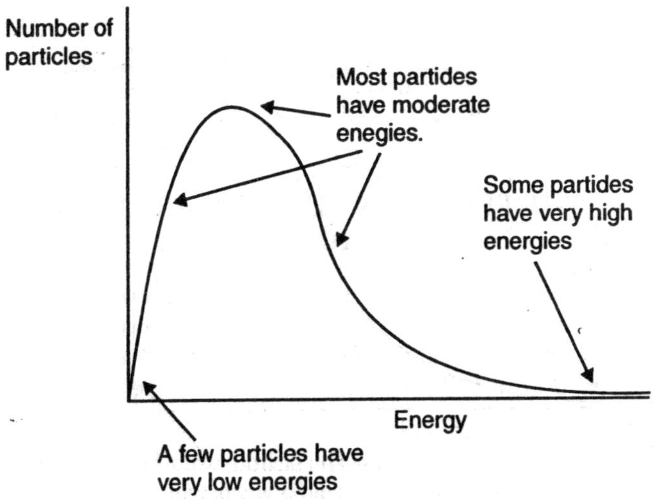

The area under the curve is a measure of the total number of particles present.

THE MAXWELL BOLTZMANN DISTRIBUTION AND ACTIVATION ENERGY

Remember that for a reaction to happen, particles must collide with energies equal to or greater than the activation energy for the reaction. We can mark the activation energy on the Maxwell-Boltzmann distribution:

Notice that the large majority of the particles don't have enough energy to react when they collide. To enable them to react we either have to change the shape of the curve, or move the activation energy further to the left. This is described on other pages.

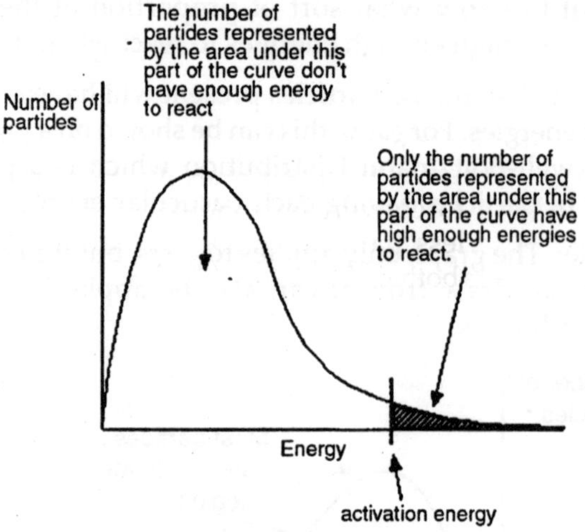

THE EFFECT OF SURFACE AREA ON REACTION RATES

Here we describes and explains the effect of changing the surface area of a solid on the rate of a reaction it is involved in. This applies to reactions involving a solid and a gas, or a solid and a liquid. It includes cases where the solid is acting as a catalyst.

The more finely divided the solid is, the faster the reaction happens. A powdered solid will *normally* produce a faster reaction than if the same mass is present as a single lump. The powdered solid has a greater surface area than the single lump.

SOME EXAMPLES

Calcium Carbonate and Hydrochloric Acid

In the lab, powdered calcium carbonate reacts much faster with dilute hydrochloric acid than if the same mass was present as lumps of marble or limestone.

$$CaCO_{3(s)} + 2HCl_{(aq)} \longrightarrow CaCl_{2(aq)} + H_2O_{(l)} + CO_{2(g)}$$

The Catalytic Decomposition of Hydrogen Peroxide

This is another familiar lab reaction. Solid manganese(IV) oxide is often used as the catalyst. Oxygen is given off much faster if the catalyst is present as a powder than as the same mass of granules.

$$2H_2O_{2(aq)} \xrightarrow{MnO_{2(s)}} 2H_2O_{(l)} + O_{2(g)}$$

CATALYTIC CONVERTERS

Catalytic converters use metals like platinum, palladium and rhodium to convert poisonous compounds in vehicle exhausts into less harmful things. For example, a reaction which removes both carbon monoxide and an oxide of nitrogen is:

$$2CO_{(g)} + 2NO_{(g)} \longrightarrow 2CO_{2(g)} + N_{2(g)}$$

Because the exhaust gases are only in contact with the catalyst for a very short time, the reactions have to be very fast. The extremely expensive metals used as the catalyst are coated as a very thin layer onto a ceramic honeycomb structure to maximise the surface area.

The Explanation

You are only going to get a reaction if the particles in the gas or liquid collide with the particles in the solid.Increasing the surface area of the solid increases the chances of collision taking place.

Imagine a reaction between magnesium metal and a dilute acid like hydrochloric acid. The reaction involves collision between magnesium atoms and hydrogen ions.

$$Mg_{(s)} + 2H^+_{(aq)} \longrightarrow Mg^{2+}_{(aq)} + H_{2(g)}$$

Increasing the number of collisions per second increases the rate of reaction.

THE EFFECT OF CONCENTRATION ON REACTION RATES

This page describes and explains the way that changing the concentration of a solution affects the rate of a reaction. Be aware that this is an introductory page only. If you are interested in orders of reaction, you will find separate pages dealing with these. You can access these via the rates of reaction menu.

For many reactions involving liquids or gases, increasing the concentration of the reactants increases the rate of reaction. In a few cases, increasing the concentration of one of the reactants may have little noticeable effect of the rate. These cases are discussed and explained further down this page.

Don't assume that if you double the concentration of one of the reactants that you will double the rate of the reaction. It may happen like that, but the relationship may well be more complicated.

Some Examples

The examples here all involve solutions. Changing the concentration of a gas is achieved by changing its pressure. This is covered on a separate page.

Zinc and Hydrochloric Acid

In the lab, zinc granules react fairly slowly with dilute hydrochloric acid, but much faster if the acid is concentrated.

$$Zn_{(s)} + 2HCl_{(aq)} \longrightarrow ZnCl_{2(aq)} + H_{2(g)}$$

The Catalytic Decomposition of Hydrogen Peroxide

Solid manganese(IV) oxide is often used as a catalyst in this reaction. Oxygen is given off much faster if the hydrogen peroxide is concentrated than if it is dilute.

$$2H_2O_{2(aq)} \xrightarrow{MnO_{2(s)}} 2H_2O_{(l)} + O_{2(g)}$$

The Reaction between Sodium Thiosulphate Solution and Hydrochloric Acid

This is a reaction which is often used to explore the relationship between concentration and rate of reaction in introductory courses (like GCSE). When a dilute acid is added to sodium thiosulphate solution, a pale yellow precipitate of sulphur is formed.

$$Na_2S_2O_{3(aq)} + 2HCl_{(aq)} \longrightarrow 2NaCl_{(aq)} + H_2O_{(l)} + S_{(s)} + SO_{2(g)}$$

As the sodium thiosulphate solution is diluted more and more, the precipitate takes longer and longer to form.

The Explanation

Cases where changing the concentration affects the rate of the reaction.This is the common case, and is easily explained.

Collisions Involving two Particles

The same argument applies whether the reaction involves collision between two different particles or two of the same particle.

In order for any reaction to happen, those particles must first collide. This is true whether both particles are in solution, or whether one is in solution and the other a solid. If the concentration is higher, the chances of collision are greater.

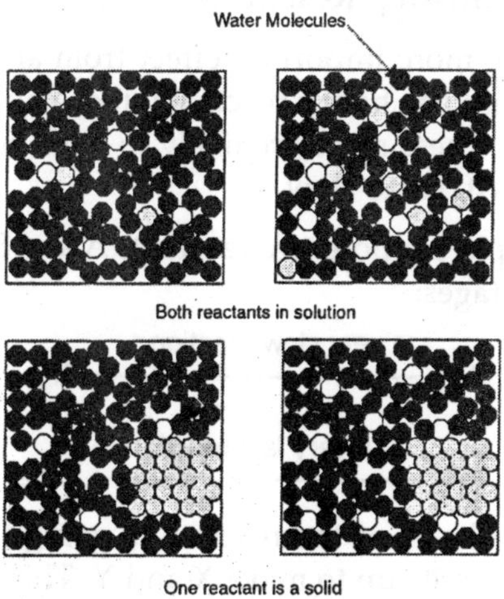

Figure. Collisions Involving two Particles

Reactions Involving only one Particle

If a reaction only involves a single particle splitting up in some way, then the number of collisions is irrelevant. What matters now is how many of the particles have enough energy to react at any one time.

Suppose that at any one time 1 in a million particles have enough energy to equal or exceed the activation energy. If you had 100 million particles, 100 of them would react. If you had 200 million particles in the same volume, 200 of them would now react. The rate of reaction has doubled by doubling the concentration.

CASES WHERE CHANGING THE CONCENTRATION DOESN'T AFFECT THE RATE OF THE REACTION

Suppose you are using a small amount of a solid catalyst in a reaction, and a high enough concentration of reactant in solution so that the catalyst surface was totally cluttered up with reacting particles.

Increasing the concentration of the solution even more can't have any effect because the catalyst is already working at its maximum capacity.

In Certain Multi-step Reactions

This is the more important effect from an A' level point of view. Suppose you have a reaction which happens in a series of small steps. These steps are likely to have widely different rates - some fast, some slow.

For example, suppose two reactants A and B react together in these two stages:

$$A_{(aq)} \xrightarrow{\text{a very slow reaction}} X_{(aq)} + Y_{(aq)}$$

$$X_{(aq)} + B_{(aq)} \xrightarrow{\text{a very fast reaction}} \text{Other products}$$

The overall rate of the reaction is going to be governed by how fast A splits up to make X and Y. This is described as the *rate determining step* of the reaction.

If you increase the concentration of A, you will increase the chances of this step happening for reasons we've looked at above.

If you increase the concentration of B, that will undoubtedly speed up the second step, but that makes hardly any difference to the overall rate. You can picture the second step as happening so fast already that as soon as any X is formed, it is immediately pounced on by B. That second reaction is already "waiting around" for the first one to happen.

The best specific examples of reactions of this type comes from organic chemistry. These involve the reaction between a tertiary halogenoalkane (alkyl halide) and a number of possible substances - including hydroxide ions. These are examples of nucleophilic substitution using a mechanism known as S_N1.

THE EFFECT OF PRESSURE ON REACTION RATES

This page describes and explains the way that changing the pressure of a gas changes the rate of a reaction.

Increasing the pressure on a reaction involving reacting gases increases the rate of reaction. Changing the pressure on a reaction which involves only solids or liquids has no effect on the rate.

An Example

In the manufacture of ammonia by the Haber Process, the rate of reaction between the hydrogen and the nitrogen is increased by the use of very high pressures.

$$N_{2(g)} + 3H_{2(g)} \rightleftharpoons 2NH_{3(g)}$$

In fact, the main reason for using high pressures is to improve the percentage of ammonia in the equilibrium mixture, but there is a useful effect on rate of reaction as well.

THE RELATIONSHIP BETWEEN PRESSURE AND CONCENTRATION

Increasing the pressure of a gas is exactly the same as increasing its concentration. If you have a given mass of gas, the way you increase its pressure is to squeeze it into a smaller volume. If you have the same mass in a smaller volume, then its concentration is higher.

You can also show this relationship mathematically if you have come across the *ideal gas equation:*

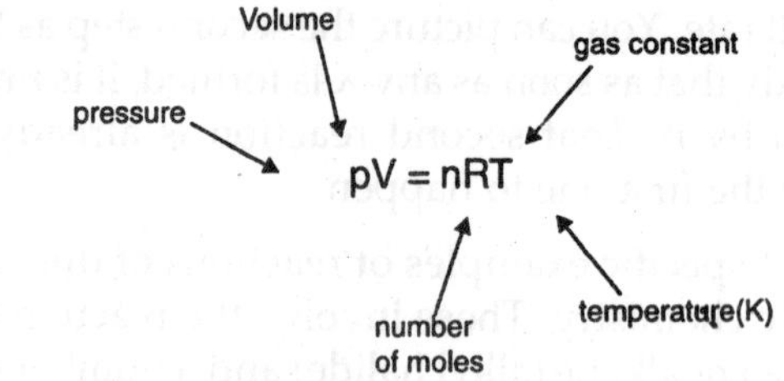

Rearranging this gives:

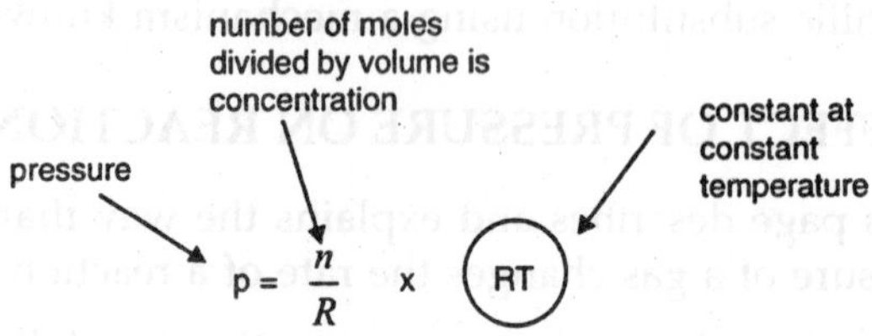

Because "RT" is constant as long as the temperature is constant, this shows that the pressure is directly proportional to the concentration. If you double one, you will also double the other.

THE EFFECT OF INCREASING THE PRESSURE ON THE RATE OF REACTION

Collisions Involving two Particles

The same argument applies whether the reaction involves collision between two different particles or two of the same particle.

In order for any reaction to happen, those particles must first collide. This is true whether both particles are in the gas state, or whether one is a gas and the other a solid. If the pressure is higher, the chances of collision are greater.

Reactions Involving only one Particle

If a reaction only involves a single particle splitting up in some way, then the number of collisions is irrelevant. What matters now is how many of the particles have enough energy to react at any one time.

Suppose that at any one time 1 in a million particles have enough energy to equal or exceed the activation energy. If you had 100 million particles, 100 of them would react. If you had 200 million particles in the same volume, 200 of them would now react. The rate of reaction has doubled by doubling the pressure.

THE EFFECT OF TEMPERATURE ON REACTION RATES

This page describes and explains the way that changing the temperature affects the rate of a reaction. It assumes that you are already familiar with basic ideas about the collision theory, and with the Maxwell-Boltzmann distribution of molecular energies in a gas.

As you increase the temperature the rate of reaction increases. As a rough approximation, for many reactions happening at around room temperature, the rate of reaction doubles for every 10°C rise in temperature.

You have to be careful not to take this too literally. It doesn't apply to all reactions. Even where it is approximately true, it may be that the rate doubles every 9°C or 11°C or whatever. The number of degrees needed to double the rate will also change gradually as the temperature increases.

Examples

Some reactions are virtually instantaneous - for example, a precipitation reaction involving the coming together of ions in solution to make an insoluble solid, or the reaction between hydrogen ions from an acid and hydroxide ions from an alkali in solution. So heating one of these won't make any noticeable difference to the rate of the reaction.

Almost any other reaction you care to name will happen faster if you heat it - either in the lab, or in industry.

THE EXPLANATION

Increasing the Collision Frequency

Particles can only react when they collide. If you heat a

substance, the particles move faster and so collide more frequently. That will speed up the rate of reaction.

That seems a fairly straightforward explanation until you look at the numbers!

It turns out that the frequency of two-particle collisions in gases is proportional to the square root of the kelvin temperature. If you increase the temperature from 293 K to 303 K (20°C to 30 and degC), you will increase the collision frequency by a factor of:

$$\sqrt{\frac{303}{293}} = 1.017$$

That's an increase of 1.7 per cent for a 10° rise. The rate of reaction will probably have doubled for that increase in temperature - in other words, an increase of about 100 per cent. The effect of increasing collision frequency on the rate of the reaction is *very* minor. The important effect is quite different.

THE KEY IMPORTANCE OF ACTIVATION ENERGY

Collisions only result in a reaction if the particles collide with enough energy to get the reaction started. This minimum energy required is called the activation energy for the reaction.

You can mark the position of activation energy on a Maxwell-Boltzmann distribution to get a diagram like this:

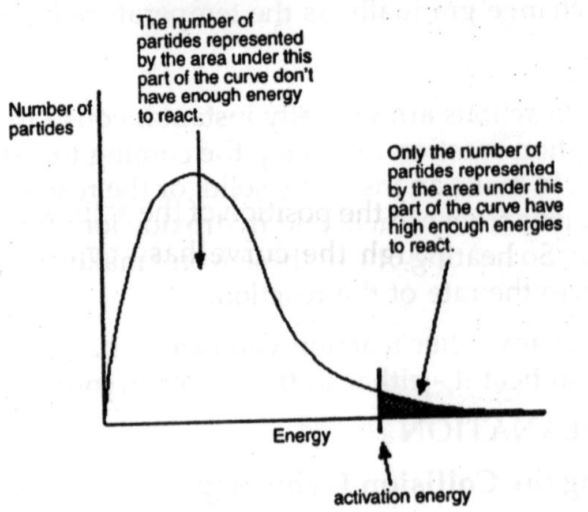

Fig. The Key Importance of Activation Energy

Only those particles represented by the area to the right of the activation energy will react when they collide. The great majority don't have enough energy, and will simply bounce apart.

To speed up the reaction, you need to increase the number of the very energetic particles - those with energies equal to or greater than the activation energy. Increasing the temperature has exactly that effect - it changes the shape of the graph.

In the next diagram, the graph labelled *T* is at the original temperature. The graph labelled *T+t* is at a higher temperature.

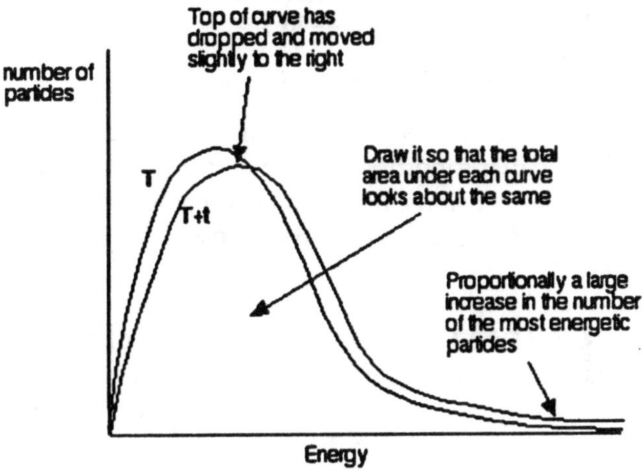

If you now mark the position of the activation energy, you can see that although the curve hasn't moved very much overall, there has been such a large increase in the number of the very energetic particles that many more now collide with enough energy to react.

Remember that the area under a curve gives a count of the number of particles. On the last diagram, the area under the higher temperature curve to the right of the activation energy looks to have at least doubled - therefore at least doubling the rate of the reaction.

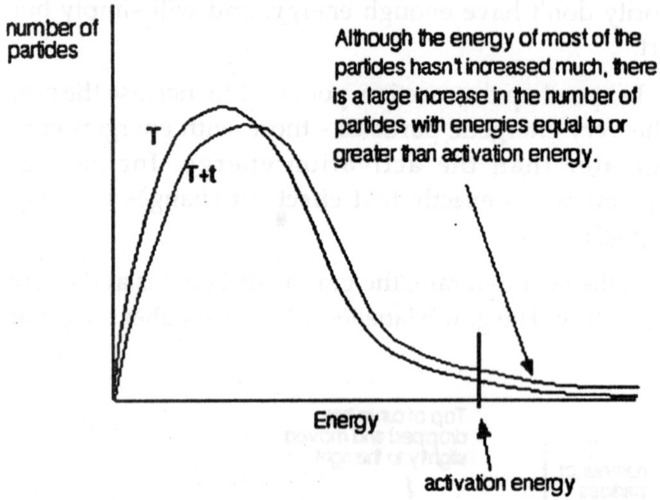

Increasing the temperature increases reaction rates because of the disproportionately large increase in the number of high energy collisions. It is only these collisions (possessing *at least* the activation energy for the reaction) which result in a reaction.

THE EFFECT OF CATALYSTS ON REACTION RATES

This page describes and explains the way that adding a catalyst affects the rate of a reaction. It assumes that you are already familiar with basic ideas about the collision theory of reaction rates, and with the Maxwell-Boltzmann distribution of molecular energies in a gas.

Note that this is only a preliminary look at catalysis as far as it affects rates of reaction. If you are looking for more detail, there is a separate section dealing with catalysts which you can access via a link at the bottom of the page.

A catalyst is a substance which speeds up a reaction, but is chemically unchanged at the end of the reaction. When the reaction has finished, you would have exactly the same mass of catalyst as you had at the beginning.

Some Examples

Some common examples which you may need for other parts of your syllabus include:

Reaction	Catalyst
Decomposition of hydrogen peroxide	manganese(IV) oxide, MnO_2
Nitration of benzene	concentrated sulphuric acid
Manufacture of ammonia by the Haber Process	iron
Conversion of SO_2 into SO_3 during the Contact Process to make sulphuric acid	vanadium(V) oxide, V_2O_5
Hydrogenation of a C=C double bond	nickel

The Explanation

THE KEY IMPORTANCE OF ACTIVATION ENERGY

Collisions only result in a reaction if the particles collide with enough energy to get the reaction started. This minimum energy required is called the activation energy for the reaction.

You can mark the position of activation energy on a Maxwell-Boltzmann distribution to get a diagram like this:

Only those particles represented by the area to the right of the activation energy will react when they collide. The great majority don't have enough energy, and will simply bounce apart.

CATALYSTS AND ACTIVATION ENERGY

To increase the rate of a reaction you need to increase the number of successful collisions. One possible way of doing this is to provide an alternative way for the reaction to happen which has a lower activation energy.

In other words, to move the activation energy on the graph like this:

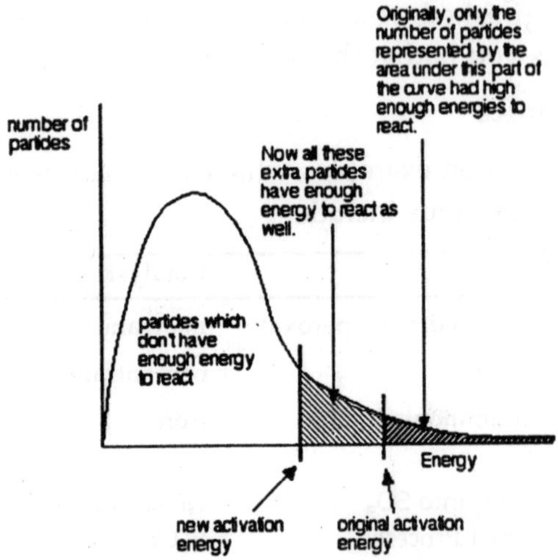

Adding a catalyst has exactly this effect on activation energy. A catalyst provides an alternative route for the reaction. That alternative route has a lower activation energy. Showing this on an energy profile:

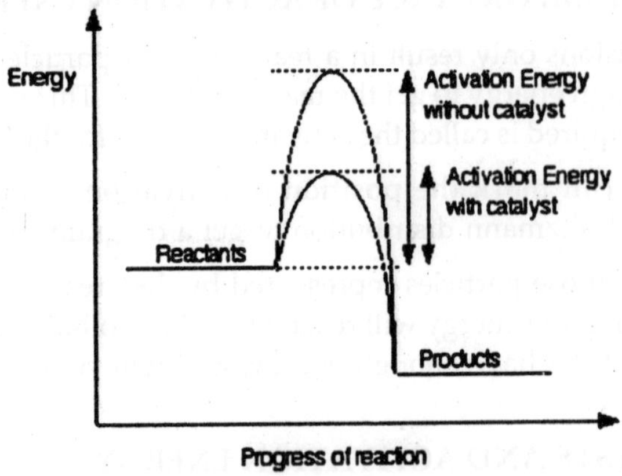

ORDERS OF REACTION AND RATE EQUATIONS

Changing the concentration of substances taking part in a reaction usually changes the rate of the reaction. A rate equation shows this effect mathematically. Orders of reaction are a part of the rate equation. This page introduces and explains the various terms you will need to know about.

RATE EQUATIONS

Measuring a Rate of Reaction

There are several simple ways of measuring a reaction rate. For example, if a gas was being given off during a reaction, you could take some measurements and work out the volume being given off per second at any particular time during the reaction.

A rate of 2 $cm^3 s^{-1}$ is obviously twice as fast as one of 1 $cm^3 s^{-1}$.

However, for this more formal and mathematical look at rates of reaction, the rate is usually measured by looking at how fast the *concentration of one of the reactants* is falling at any one time.

For example, suppose you had a reaction between two substances A and B. Assume that at least one of them is in a form where it is sensible to measure its concentration - for example, in solution or as a gas.

$$A + B \rightarrow product$$

For this reaction you could measure the rate of the reaction by finding out how fast the concentration of, say, A was falling per second.

You might, for example, find that at the beginning of the reaction, its concentration was falling at a rate of 0.0040 mol $dm^{-3} s^{-1}$.

This means that every second the concentration of A was falling by 0.0040 moles per cubic decimetre. This rate will decrease during the reaction as A gets used up.

For the purposes of rate equations and orders of reaction, the rate of a reaction is measured in terms of how fast the concentration of one of the reactants is falling. Its units are mol $dm^{-3} s^{-1}$.

ORDERS OF REACTION

Orders of reaction are always found by doing experiments. You can't deduce anything about the order of a reaction just by looking at the equation for the reaction.

So let's suppose that you have done some experiments to find out what happens to the rate of a reaction as the concentration of one of the reactants, A, changes. Some of the simple things that you might find are:

One Possibility: The Rate of Reaction is Proportional to the Concentration of A

That means that if you double the concentration of A, the rate doubles as well. If you increase the concentration of A by a factor of 4, the rate goes up 4 times as well.

You can express this using symbols as:

proportionality sign

$$\text{rate} \;\alpha\; [A]$$

square brackets mean "concentration in mol dm^{-3}".

Writing a formula in square brackets is a standard way of showing a concentration measured in moles per cubic decimetre (litre).

constant known as the "rate constant"

$$\text{rate} = k[A]$$

You can also write this by getting rid of the proportionality sign and introducing a constant, k.

Another Possibility: The Rate of Reaction is Proportional to The Square of The Concentration of A

This means that if you doubled the concentration of A, the rate would go up 4 times (2^2). If you tripled the concentration of A, the rate would increase 9 times (3^2). In symbol terms:

$$\text{Rate} \ \alpha \ [A]^2$$

$$\text{rate} = k[A]^2$$

By doing experiments involving a reaction between A and B, you would find that the rate of the reaction was related to the concentrations of A and B in this way:

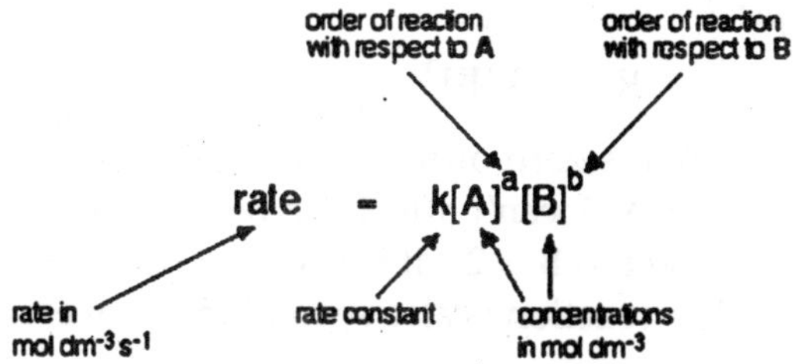

This is called the *rate equation* for the reaction.

The concentrations of A and B have to be raised to some power to show how they affect the rate of the reaction. These powers are called the *orders of reaction* with respect to A and B.

For UK A' level purposes, the orders of reaction you are likely to meet will be 0, 1 or 2. But other values are possible including fractional ones like 1.53, for example.

If the order of reaction with respect to A is 0 (zero), this means that the concentration of A doesn't affect the rate of reaction. Mathematically, any number raised to the power of zero (x^0) is equal to 1. That means that that particular term disappears from the rate equation.

The *overall order of the reaction* is found by adding up the individual orders. For example, if the reaction is first order with respect to both A and B ($a = 1$ and $b = 1$), the overall order is 2. We call this an overall second order reaction.

Some Examples

Each of these examples involves a reaction between A and B, and each rate equation comes from doing some experiments to find out how the concentrations of A and B affect the rate of reaction.

Example 1:

$$\text{Rate} = k[A][B]$$

In this case, the order of reaction with respect to both A and B is 1. The overall order of reaction is 2– found by adding up the individual orders.

Example 2:

$$\text{Rate} = k[B]^2$$

This reaction is zero order with respect to A because the concentration of A doesn't affect the rate of the reaction. The order with respect to B is 2 - it's a second order reaction with respect to B. The reaction is also second order overall (because 0 + 2 = 2).

Example 3:

$$\text{Rate} = k[A]$$

This reaction is first order with respect to A and zero order with respect to B, because the concentration of B doesn't affect the rate of the reaction. The reaction is first order overall (because 1 + 0 = 1).

It doesn't matter how many reactants there are. The concentration of each reactant will occur in the rate equation, raised to some power. Those powers are the individual orders of reaction. The overall order of the reaction is found by adding them all up.

THE RATE CONSTANT

Surprisingly, the rate constant isn't actually a true constant! It varies, for example, if you change the temperature of the reaction, add a catalyst, or change the catalyst.

The rate constant is constant for a given reaction only if all you are changing is the concentration of the reactants. You will find more about the effect of temperature and catalysts on the rate constant on another page.

CALCULATIONS INVOLVING ORDERS OF REACTION

You will almost certainly have to be able to calculate orders of reaction and rate constants from given data or from your own experiments.

There are all sorts of ways of doing these sums, and it is important that you practice the methods that your syllabus wants. Check your syllabus and past exam papers to see what sort of examples you need to be able to work out.

ORDERS OF REACTION AND MECHANISMS

This page looks at the relationship between orders of reaction and mechanisms in some simple cases. It explores what a mechanism is, and the idea of a rate determining step. It also explains the difference between the sometimes confusing terms "order of reaction" and "molecularity of reaction".

REACTION MECHANISMS

In any chemical change, some bonds are broken and new ones are made. Quite often, these changes are too complicated to happen in one simple stage. Instead, the reaction may involve a series of small changes one after the other.

A reaction mechanism describes the one or more steps involved in the reaction in a way which makes it clear exactly how the various bonds are broken and made. The following example comes from organic chemistry. It doesn't matter in the least if it is unfamiliar to you!

This is a reaction between 2-bromo-2-methylpropane and the hydroxide ions from sodium hydroxide solution:

$$(CH_3)_3CBr + OH^- \longrightarrow (CH_3)_3COH + Br^-$$

The overall reaction replaces the bromine atom in the organic compound by an OH group.

The first thing that happens is that the carbon-bromine bond in a small proportion of the organic compound breaks to give ions:

Carbon-bromine bonds are reasonably strong, so this is a slow change. If the ions hit each other again, the covalent bond will reform. The curly arrow in the equation shows the movement of a pair of electrons.

If there is a high concentration of hydroxide ions present, the positive ion stands a high chance of hitting one of those. This step of the overall reaction will be very fast. A new covalent bond is made between the carbon and the oxygen, using one of the lone pairs on the oxygen atom.

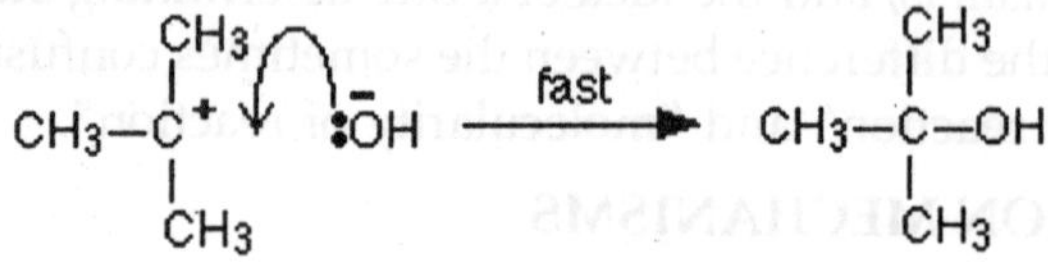

Because carbon-oxygen bonds are strong, once the OH group has attached to the carbon atom, it tends to stay attached.

The mechanism shows that the reaction takes place in two steps and describes exactly how those steps happen in terms of bonds being broken or made. It also shows that the steps have different rates of reaction - one slow and one fast.

THE RATE DETERMINING STEP

The overall rate of a reaction (the one which you would measure if you did some experiments) is controlled by the rate of the slowest step. In the example above, the hydroxide ion can't combine with the positive ion until that positive ion has been formed. The second step is in a sense waiting around for the first slow step to happen. The slow step of a reaction is known as the *rate determining step*.

As long as there is a lot of difference between the rates of the various steps, when you measure the rate of a reaction, you are actually measuring the rate of the rate determining step.

REACTION MECHANISMS AND ORDERS OF REACTION

The examples we use at this level are the very simple ones where the orders of reaction with respect to the various substances taking part are 0, 1 or 2. These tend to have the slow step of the reaction happening before any fast step(s).

To try to explain how fractional orders of reaction can arise is beyond the scope of UK A′ level courses.

Example 1

Here is the mechanism we have already looked at. How do we know that it works like this?

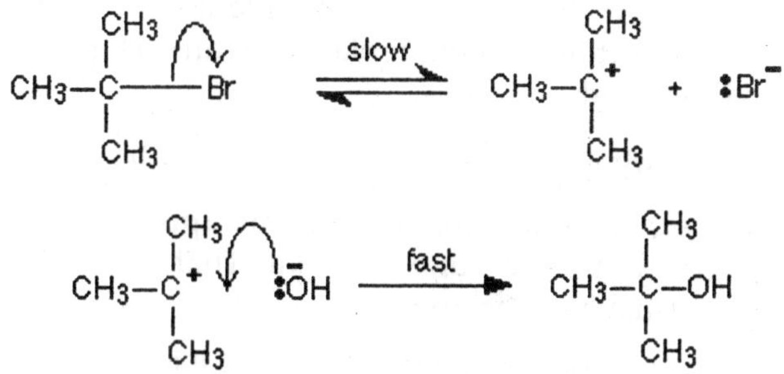

By doing rate of reaction experiments, you find this rate equation:

$$\text{Rate} = k[(CH_3)_3C\text{–}Br]$$

The reaction is first order with respect to the organic compound, and zero order with respect to the hydroxide ions. The concentration of the hydroxide ions isn't affecting the overall rate of the reaction.

If the hydroxide ions were taking part in the slow step of the reaction, increasing their concentration would speed the reaction up. Since their concentration doesn't seem to matter, they must be taking part in a later fast step.

Increasing the concentration of the hydroxide ions will speed up the fast step, but that won't have a noticeable effect on the overall rate of the reaction. That is governed by the speed of the slow step.

In a simple case like this, where the slow step of the reaction is the first step, the rate equation tells you what is taking part in that slow step.In this case, the reaction is first order with respect to the organic molecule - and that's all.

This gives you a starting point for working out a possible mechanism. Having come up with a mechanism, you would need to find more evidence to confirm it. For example, in this case you might try to detect the presence of the positive ion that is formed in the first step.

Example 2

At first sight this reaction seems identical with the last one. A bromine atom is being replaced by an OH group in an organic compound.

$$CH_3CH_2Br + OH^- \longrightarrow CH_3CH_2OH + Br^-$$

However, the rate equation for this apparently similar reaction turns out to be quite different. That means that the mechanism must be different.

$$\text{rate} = k[CH_3CH_2 - Br][OH^-]$$

The reaction this time is first order with respect to both the organic compound and the hydroxide ions. Both of these must be taking part in the slow step of the reaction. The reaction must happen by a straightforward collision between them.

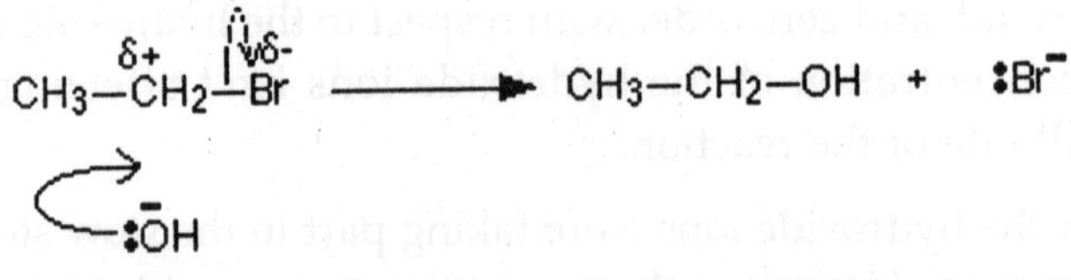

The carbon atom which is hit by the hydroxide ion has a slight positive charge on it and the bromine a slight negative one because of the difference in their electronegativities.

As the hydroxide ion approaches, the bromine is pushed off in one smooth action.

MOLECULARITY OF A REACTION

You may come across an older term known as the *molecularity* of a reaction. This has largely dropped out of UK A' level syllabuses, but if you meet it, it is important that you understand the difference between this and the *order* of a reaction. The terms were sometimes used carelessly as if they mean the same thing - they don't!

ORDER OF REACTION

The important thing to realise is that this is something which can only be found by doing experiments. It gives you information about which concentrations affect the rate of the reaction. You cannot look at an equation for a reaction and deduce what the order of the reaction is going to be - you have to do some practical work!

Having found the order of the reaction experimentally, you *may* be able to make suggestions about the mechanism for the reaction - at least in simple cases.

This starts at the other end! If you know the mechanism for a reaction, you can write down equations for a series of steps which make it up. Each of those steps has a molecularity.

The molecularity of a step simply counts the number of species (molecules, ions, atoms or free radicals) taking part in that step. For example, going back to the mechanisms we've been looking at:

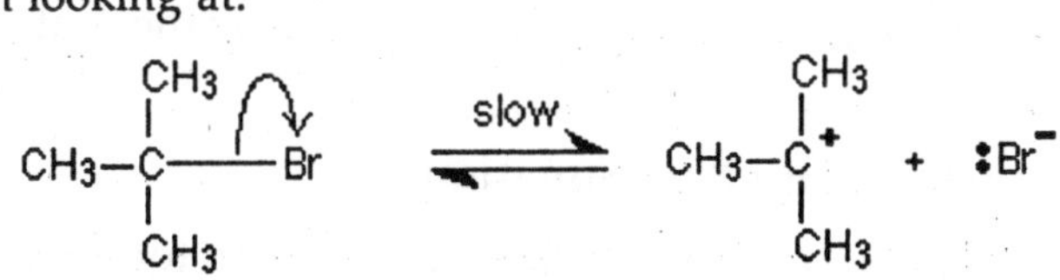

This step involves a single molecule breaking into ions. Because only one species is involved in the reaction, it has a molecularity of 1. It could be described as *unimolecular*.

The second step of this mechanism, involves two ions reacting together.

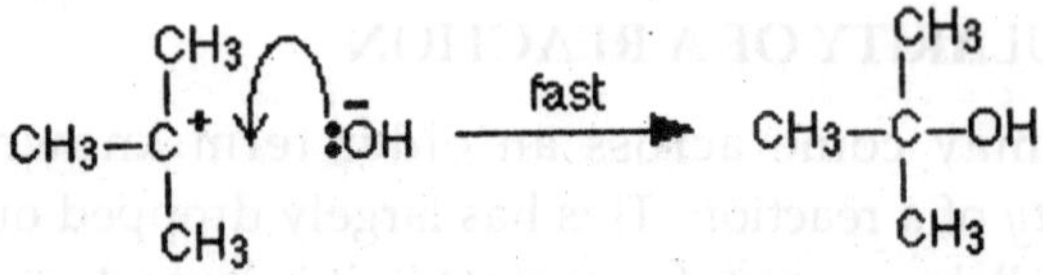

This step has a molecularity of 2 - a *bimolecular* reaction.

The other reaction we looked at happened in a single step:

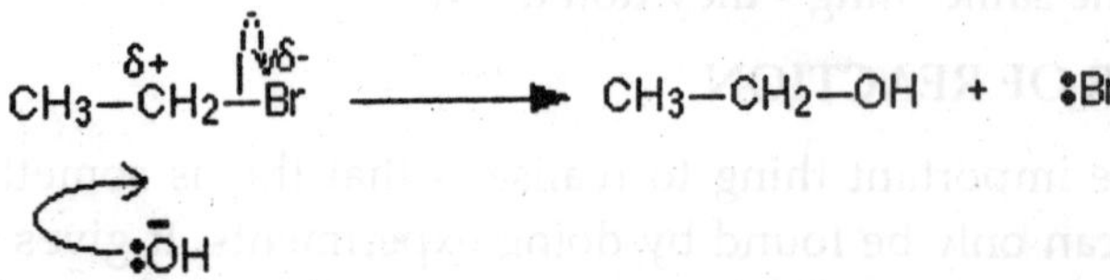

Because of the two species involved (one molecule and one ion), this reaction is also bimolecular.

Unless an *overall* reaction happens in one step (like this last one), you can't assign it a molecularity. You have to know the mechanism, and then each individual step has its own molecularity.

There's nothing the least bit complicated about the term *molecularity*. The only confusion is that you may sometimes find it used as if it meant the same as *order*. It doesn't!

RATE CONSTANTS AND THE ARRHENIUS EQUATION

This page looks at the way that rate constants vary with temperature and activation energy as shown by the Arrhenius equation.

Rate Constants and Rate Equations

You will remember that the rate equation for a reaction between two substances *A* and *B* looks like this:

The rate equation shows the effect of changing the concentrations of the reactants on the rate of the reaction. What about all the other things (like temperature and catalysts, for example) which also change rates of reaction? Where do these fit into this equation?

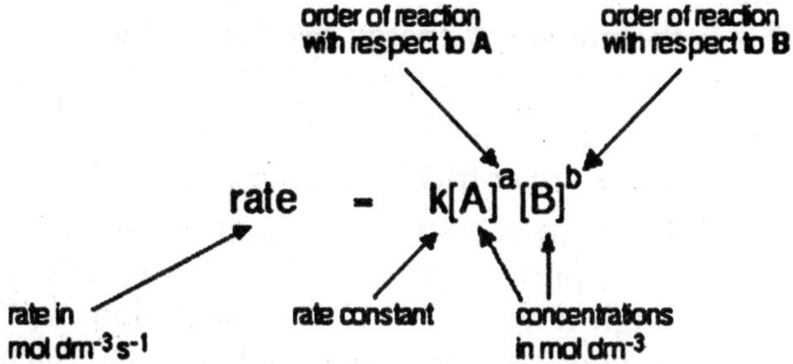

These are all included in the so-called *rate constant* - which is only actually constant if all you are changing is the concentration of the reactants. If you change the temperature or the catalyst, for example, the rate constant changes.

This is shown mathematically in the Arrhenius equation.

THE ARRHENIUS EQUATION

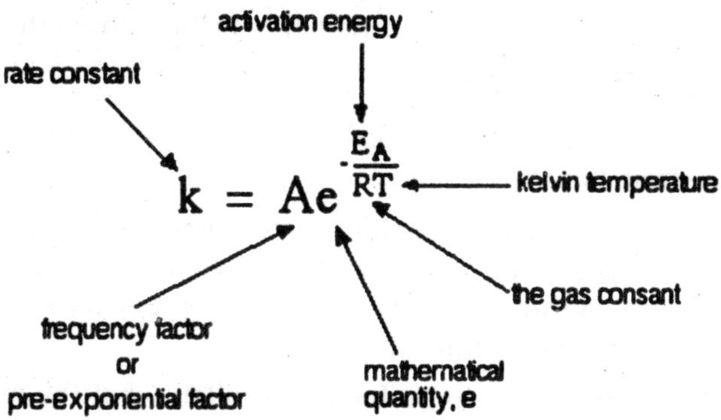

Starting with the easy ones . . .

Temperature, T

To fit into the equation, this has to be meaured in kelvin.

The gas constant, R

This is a constant which comes from an equation, pV=nRT, which relates the pressure, volume and temperature of a particular number of moles of gas. It turns up in all sorts of unlikely places!

Activation energy, E_A

This is the minimum energy needed for the reaction to occur. To fit this into the equation, it has to be expressed in joules per mole - not in kJ mol^{-1}

And then the rather trickier ones . . .

This has a value of 2.71828 . . . and is a mathematical number, a bit like pi. You don't need to worry exactly what it means, although if you have to do calculations with the Arrhenius equation, you may have to find it on your calculator. You should find an e^x button - probably on the same key as "ln".

The expression, $e^{-(E_A/RT)}$

For reasons that are beyond the scope of any course at this level, this expression counts the fraction of the molecules present in a gas which have energies equal to or in excess of activation energy at a particular temperature. You will find a simple calculation associated with this further down the page.

The frequency factor, A

You may also find this called the pre-exponential factor or the steric factor.

A is a term which includes factors like the frequency of collisions and their orientation. It varies slightly with temperature, although not much. It is often taken as constant across small temperature ranges.

By this time you've probably forgotten what the original Arrhenius equation looked like! Here it is again:

$$k = Ae^{-\frac{E_A}{RT}}$$

You may also come across it in a different form created by a mathematical operation on the standard one:

$$\ln k = \ln A - \frac{E_A}{RT}$$

"ln" is a form of logarithm. Don't worry about what it means. If you need to use this equation, just find the "ln" button on your calculator.

Using the Arrhenius Equation

You can use the Arrhenius equation to show the effect of a change of temperature on the rate constant - and therefore on the rate of the reaction. If the rate constant doubles, for example, so also will the rate of the reaction. Look back at the rate equation at the top of this page if you aren't sure why that is.

What happens if you increase the temperature by 10°C from, say, 20°C to 30°C (293 K to 303 K)?

The frequency factor, A, in the equation is approximately constant for such a small temperature change. We need to look at how $e^{-(E_A / RT)}$ changes - the fraction of molecules with energies equal to or in excess of the activation energy.

Let's assume an activation energy of 50 kJ mol^{-1}. In the equation, we have to write that as 50000 J mol^{-1}. The value of the gas constant, R, is 8.31 J K^{-1} mol^{-1}.

At 20°C (293 K) the value of the fraction is:

$$e^{-\frac{E_A}{RT}} = e^{\frac{50000}{8.31 \times 293}}$$

$$= 1.21 \times 10^{-9}$$

By raising the temperature just a little bit (to 303 K), this increases:

$$e^{-\frac{E_A}{RT}} = e^{-\frac{50000}{8.31 \times 303}}$$

$$= 2.38 \times 10^{-9}$$

You can see that the fraction of the molecules able to react has almost doubled by increasing the temperature by 10°C. That causes the rate of reaction to almost double. This is the value in the rule-of-thumb often used in simple rate of reaction work

The Effect of a Catalyst

A catalyst will provide a route for the reaction with a lower activation energy. Suppose in the presence of a catalyst that the activation energy falls to 25 kJ mol^{-1}. Redoing the calculation at 293 K:

$$e^{-\frac{E_A}{RT}} = e^{-\frac{25000}{8.31 \times 293}}$$

$$= 3.47 \times 10^{-5}$$

If you compare that with the corresponding value where the activation energy was 50 kJ mol^{-1}, you will see that there has been a massive increase in the fraction of the molecules which are able to react. There are almost 30000 times more molecules which can react in the presence of the catalyst compared to having no catalyst (using our assumptions about the activation energies).

It's no wonder catalysts speed up reactions!

Other Calculations Involving the Arrhenius Equation

If you have values for the rate of reaction or for the rate constant at different temperatures, you can use these to work out the activation energy of the reaction. Only one UK A' level Exam Board expects you to be able to do these calculations. They are included in my chemistry calculations book, and I can't repeat the material on this site.

The simplest collisional model for the rate constants in bimolecular elementary reactions, e.g.,

$$A + B \rightarrow \text{Products},$$

is to assume that reaction occurs every time an A and B molecule collide. Thus the rate becomes:

$$\text{Rate} = Z_{A,B} \qquad (1)$$

where $Z_{A,B}$ is the number of AB collisions occurring per unit time, per unit volume of the system. The kinetic theory of gases gives

$$Z_{A,B} = \pi(R_A + R_B)^{2\bar{C}} AB^n A^n B, \qquad (2)$$

where R_A and R_B are the van der Waals radii of A and B, respectively, n_A and n_B are their number densities, and $\bar{c}_{AB}$ is the mean relative speed of A with respect to B; i.e.,

$$\bar{c}_{AB} \equiv \left(\frac{8k_BT}{\pi\mu_{AB}}\right)^{1/2} \tag{3}$$

In Eq. (3), k_B is Boltzmann's constant $\mu_{AB} \equiv m_A m_B/(m_A + m_B)$ and is the AB reduced mass (denotes the mass of A or B).

Equation (2) is obtained by considering how many B molecules will lie in the cylinder swept out by an A molecule in time dt. The radius of the cylinder is the maximum distance and *A* and *B* can be in order to collide, i.e., $R_A + R_B$. Thus the volume of the cylinder for molecules moving with relative speed is

$$V(v) \equiv \pi(RA + RB)^2\, vdt. \tag{4}$$

The number of B molecules in the cylinder with relative speed v to $v + dv$ is $V(n)n_B f(v)dv$, where

$$f(v) = 4\pi v^2\left(\frac{\mu_{AB}}{2\pi k_BT}\right)^{3/2} e^{-\frac{\mu_{AB}v^2}{2k_BT}} \tag{5}$$

is the Maxwell-Boltzmann speed distribution function governing the relative motion of B with respect to A. Finally, our expression for the rate is obtained by adding (integrating) the contributions for the different relative velocities, multiplying by the number of A molecules per unit volume and dividing out the factor of dt; i.e,

$$\text{Rate} = n_A n_B \pi(R_A + R_B)^2 \int_0^\infty dv f(v) v, \tag{6}$$

which gives the expression presented above.

As was mentioned in class, Eqs. (1) - (3) in general do not correctly describe the temperature dependence of rate constants since the activation energy does not appear. In order to correct for this, we will modify our initial hypothesis; specifically, now assume that only those molecules whose relative speed is greater than v_{min} can react. In this case, Eq. (6) becomes:

$$\text{Rate} = n_A n_B \pi (R_A + R_B)^2 \int_{v_{\min}}^{\infty} dv f(v) v, \qquad (7)$$

which upon evaluating the integral gives

$$\text{Rate} = \pi (R_A + R_B)^2 \bar{c}_{AB} e^{-E_A/k_B T} \left(1 + \frac{E_A}{k_B T}\right) n_A n_B,$$

where $E_A \equiv \mu_{AB} v_{\min}^2 / 2$ is the activation energy. This expression has the correct Arrhenius form except for the extra nonexponential factors containing temperature. However, compared with the exponential factor, these usually do not change very rapidly in temperature and can approximately be treated as constant. Thus we have derived an approximate expression for the bimolecular rate constant which has at least some of the qualitative features of those found in gas reactions.

Chapter 2

Chemical Equilibrium

This chapter looks at the basic ideas underpinning the idea of a chemical equilibrium. It talks about reversible reactions and how they behave if the system is closed. This leads to the idea of a dynamic equilibrium, and what the common term "position of equilibrium" means.

REVERSIBLE REACTIONS

A reversible reaction is one which can be made to go in either direction depending on the conditions.

If you pass steam over hot iron the steam reacts with the iron to produce a black, magnetic oxide of iron called triiron tetroxide, Fe_3O_4.

$$3Fe_{(s)} + 4H_2O_{(g)} \longrightarrow Fe_3O_{4(s)} + 4H_{2(g)}$$

The hydrogen produced in the reaction is swept away by the stream of steam.

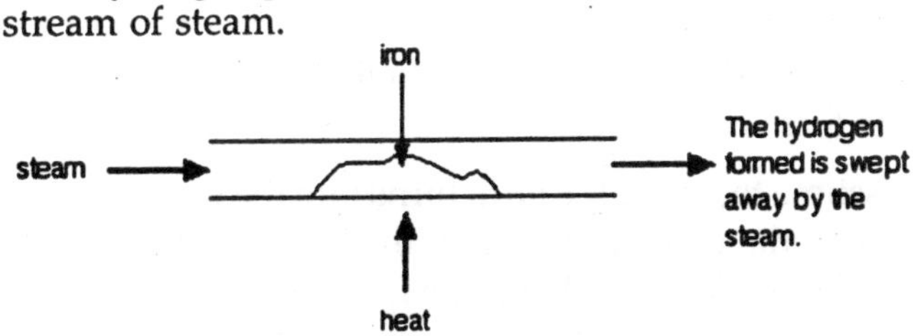

Under different conditions, the products of this reaction will also react together. Hydrogen passed over hot triiron tetroxide reduces it to iron. Steam is also produced.

This time the steam produced in the reaction is swept away by the stream of hydrogen.

These reactions are reversible, but under the conditions normally used, they become one-way reactions. The products aren't left in contact with each other, so the reverse reaction can't happen.

REVERSIBLE REACTIONS HAPPENING IN A CLOSED SYSTEM

A *closed system* is one in which no substances are either added to the system or lost from it. Energy can, however, be transferred in or out at will.

In the example we've been looking at, you would have to imagine iron being heated in steam in a closed container. Heat is being added to the system, but none of the substances in the reaction can escape. The system is closed.

As the triiron tetroxide and hydrogen start to be formed, they will also react again to give the original iron and steam. So, if you analysed the mixture after a while, what would you find?

You would find that you had established what is known as a *dynamic equilibrium*. To explain what that means, we are going to use a much simpler example . . .

DYNAMIC EQUILIBRIA

Imagine a substance which can exist in two forms - a blue form or an orange form - and that each form can react to give the other one. We are going to let them react in a closed system. Neither form can escape.

Assume that the blue form turns into the orange one much faster than the other way round. In any given time, these are the chances of the two changes happening:

■: 3 in 6 chance of changing to ■

■: 1 in 6 chance of changing to ■

You can simulate this very easily with some coloured paper cut up into small pieces (a different colour on each side), and a dice.

The following are the real results of a "reaction" I did myself. I started with 16 blue squares and looked at each one in turn and decided whether it should change colour by throwing a dice.

A blue square was turned into an orange square (the bit of paper was turned over!) if I threw a 4, 5 or 6

An orange square was turned into a blue square only if I threw a 6 while I was looking at that particular square.

Once I had looked at all 16 squares, I started the process all over again - but obviously with a different starting pattern. The diagrams show the results of doing this 11 times (plus the original 16 blue squares).

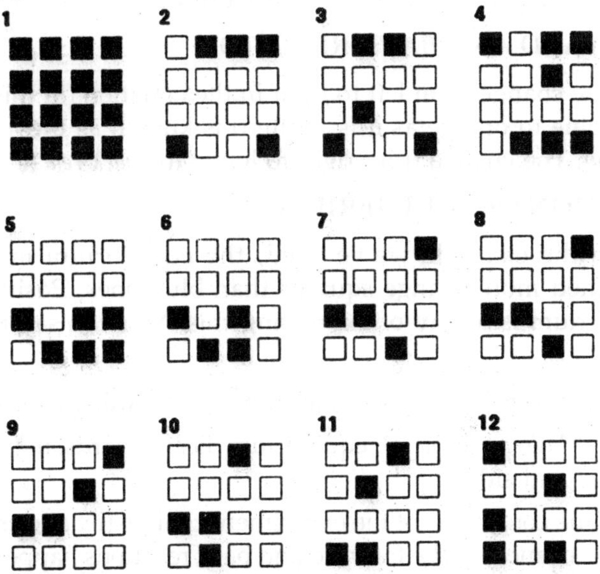

You can see that the "reaction" is continuing all the time. The exact pattern of orange and blue is constantly changing. However, the overall numbers of orange and of blue squares remain remarkably constant - most commonly, 12 orange ones to 4 blue ones.

The reaction has reached equilibrium in the sense that there is no further change in the numbers of blue and orange squares. However, the reaction is still continuing. For every orange square that turns blue, somewhere in the mixture it is replaced by a blue square turning orange.

This is known as a *dynamic equilibrium*. The word *dynamic* shows that the reaction is still continuing.

You can show dynamic equilibrium in an equation for a reaction by the use of special arrows. In the present case, you would write it as:

$$\blacksquare \rightleftharpoons \square$$

It is important to realise that this doesn't just mean that the reaction is reversible. It means that you have a reversible reaction in a state of dynamic equilibrium.

FORWARD AND BACK REACTION

The change from left to right in the equation (in this case from blue to orange as it is written) is known as the *forward reaction*. The change from right to left is the *back reaction*.

POSITION OF EQUILIBRIUM

In the example we've used, the equilibrium mixture contained more orange squares than blue ones. Position of equilibrium is a way of expressing this. You can say things like:

- "The position of equilibrium lies towards the orange."
- "The position of equilibrium lies towards the right-hand side."

If the conditions of the experiment change (by altering the relative chances of the forward and back reactions happening), the composition of the equilibrium mixture will also change.

For example, if changing the conditions produced more blue in the equilibrium mixture, you would say "The position of equilibrium has moved to the left" or "The position of equilibrium has moved towards the blue".

What happens if you started the reaction with orange squares rather than blue ones, but kept the chances of each change happening the same as in the first example? This is the result of my "reaction".

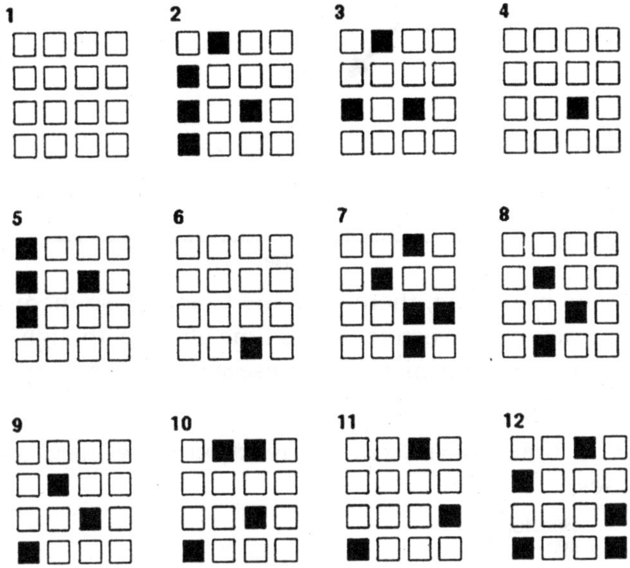

Once again, you can see that exactly the same position of equilibrium is being established as when we started with the blue squares. You get exactly the same equilibrium mixture irrespective of which side of the equation you start from - provided the conditions are the same in both cases

This is the equation for a general reaction which has reached dynamic equilibrium:

$$A + B \rightleftharpoons C + D$$

How did it get to that state? Let's assume that we started with A and B.

At the beginning of the reaction, the concentrations of A and B were at their maximum. That means that the rate of the reaction was at its fastest.

As A and B react, their concentrations fall. That means that they are less likely to collide and react, and so the rate of the forward reaction falls as time goes on.

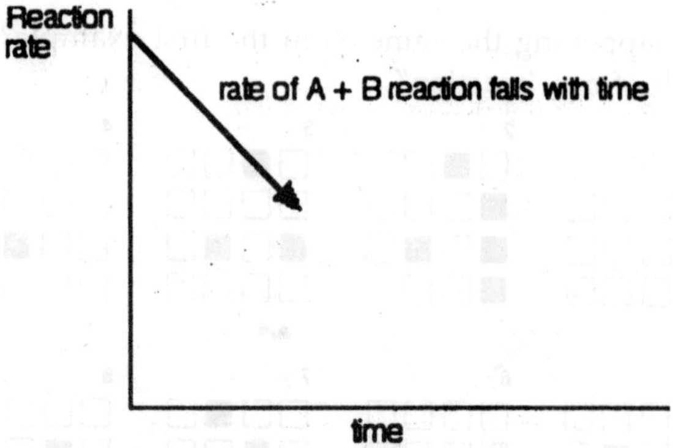

In the beginning, there isn't any C and D, so there can't be any reaction between them. As time goes on, though, their concentrations in the mixture increase and they are more likely to collide and react.

With time, the rate of the reaction between C and D increases:

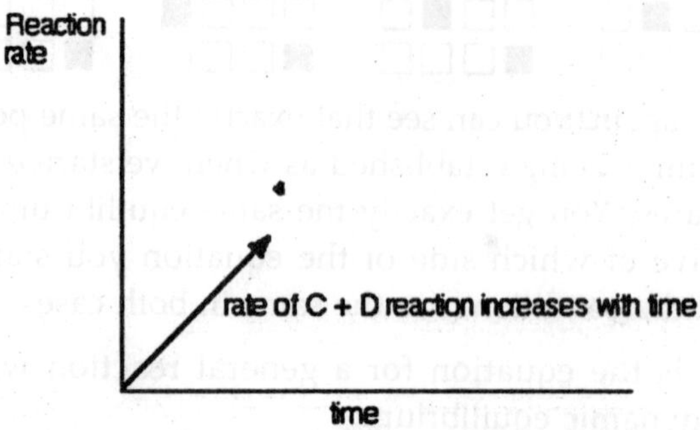

Eventually, the rates of the two reactions will become equal. A and B will be converting into C and D at exactly the same rate as C and D convert back into A and B again.

At this point there won't be any further change in the amounts of A, B, C and D in the mixture. As fast as something is being removed, it is being replaced again by the reverse reaction. We have reached a position of dynamic equilibrium.

A dynamic equilibrium occurs when you have a reversible reaction in a closed system. Nothing can be added to the system or taken away from it apart from energy.

At equilibrium, the quantities of everything present in the mixture remain constant, although the reactions are still continuing. This is because the rates of the forward and the back reactions are equal.

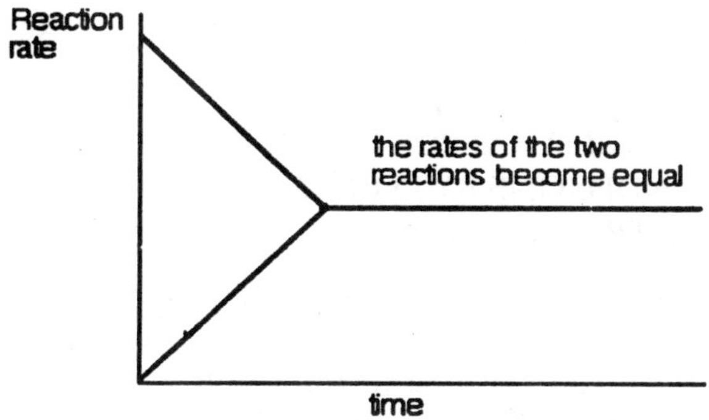

If you change the conditions in a way which changes the relative rates of the forward and back reactions you will change the position of equilibrium - in other words, change the proportions of the various substances present in the equilibrium mixture. This is explored in detail on other pages in this equilibrium section.

LE CHATELIER'S PRINCIPLE

This page looks at Le Chatelier's Principle and explains how to apply it to reactions in a state of dynamic equilibrium. It covers changes to the position of equilibrium if you change concentration, pressure or temperature. It also explains very briefly why catalysts have no effect on the position of equilibrium.

It is important in understanding everything on this page to realise that Le Chatelier's Principle is no more than a useful guide to help you work out what happens when you change the conditions in a reaction in dynamic equilibrium. *It doesn't explain anything.* I'll keep coming back to that point!

A STATEMENT OF LE CHATELIER'S PRINCIPLE

If a dynamic equilibrium is disturbed by changing the conditions, the position of equilibrium moves to counteract the change.

$$A + 2B \rightleftharpoons C + D$$

USING LE CHATELIER'S PRINCIPLE WITH A CHANGE OF CONCENTRATION

Suppose you have an equilibrium established between four substances A, B, C and D.

WHAT WOULD HAPPEN IF YOU CHANGED THE CONDITIONS BY INCREASING THE CONCENTRATION OF A

According to Le Chatelier, the position of equilibrium will move in such a way as to counteract the change. That means that the position of equilibrium will move so that the concentration of A decreases again - by reacting it with B and turning it into C + D. The position of equilibrium moves to the right.

$$A + 2B \rightleftharpoons C + D$$

$$\longrightarrow$$

The position of equilibrium moves to the right if you increase the concentration of A.

This is a useful way of converting the maximum possible amount of B into C and D. You might use it if, for example, B was a relatively expensive material whereas A was cheap and plentiful.

WHAT WOULD HAPPEN IF YOU CHANGED THE CONDITIONS BY DECREASING THE CONCENTRATION OF A

According to Le Chatelier, the position of equilibrium will move so that the concentration of A increases again. That means that more C and D will react to replace the A that has been removed. The position of equilibrium moves to the left.

$$A + 2B \rightleftharpoons C + D$$

$$\longleftarrow$$

The position of equilibrium moves to the left if you decrease the concentration of A.

This is esssentially what happens if you remove one of the products of the reaction as soon as it is formed. If, for example, you removed C as soon as it was formed, the position of equilibrium would move to the right to replace it. If you kept on removing it, the equilibrium position would keep on moving rightwards - turning this into a one-way reaction.

This isn't in any way an *explanation* of why the position of equilibrium moves in the ways described. All Le Chatelier's Principle gives you is a quick way of working out what happens.

USING LE CHATELIER'S PRINCIPLE WITH A CHANGE OF PRESSURE

This only applies to reactions involving gases:

$$A_{(g)} + 2B_{(g)} \rightleftharpoons C_{(g)} + D_{(g)}$$

WHAT WOULD HAPPEN IF YOU CHANGED THE CONDITIONS BY INCREASING THE PRESSURE

According to Le Chatelier, the position of equilibrium will move in such a way as to counteract the change. That means that the position of equilibrium will move so that the pressure is reduced again.

$$A_{(g)} + 2B_{(g)} \rightleftharpoons C_{(g)} + D_{(g)}$$

$$\longleftarrow$$

The position of equilibrium moves to the left
if you decrease the pressure on the reaction

Pressure is caused by gas molecules hitting the sides of their container. The more molecules you have in the container, the higher the pressure will be. The system can reduce the pressure by reacting in such a way as to produce fewer molecules.

$$A_{(g)} + 2B_{(g)} \rightleftharpoons C_{(g)} + D_{(g)}$$

$$\longleftarrow$$

The position of equilibrium moves to the right
if you decrease the pressure on the reaction

In this case, there are 3 molecules on the left-hand side of the equation, but only 2 on the right. By forming more C and D, the system causes the pressure to reduce.

Increasing the pressure on a gas reaction shifts the position of equilibrium towards the side with fewer molecules.

WHAT WOULD HAPPEN IF YOU CHANGED THE CONDITIONS BY DECREASING THE PRESSURE

The equilibrium will move in such a way that the pressure increases again. It can do that by producing more molecules. In this case, the position of equilibrium will move towards the left-hand side of the reaction.

WHAT HAPPENS IF THERE ARE THE SAME NUMBER OF MOLECULES ON BOTH SIDES OF THE EQUILIBRIUM REACTION

In this case, increasing the pressure has no effect whatsoever on the position of the equilibrium. Because you have the same numbers of molecules on both sides, the equilibrium can't move in any way that will reduce the pressure again.

Again, this isn't an *explanation* of why the position of equilibrium moves in the ways described. You will find a rather mathematical treatment of the explanation by following the link below.

USING LE CHATELIER'S PRINCIPLE WITH A CHANGE OF TEMPERATURE

For this, you need to know whether heat is given out or absorbed during the reaction. Assume that our forward reaction is exothermic (heat is evolved):

$$A + 2B \rightleftharpoons C + D \quad \Delta H = -250 \text{ kJ mol}^{-1}$$

250kj is evolved when A and B react completely to give C and D.

$$\longrightarrow$$

$$A + 2B \rightleftharpoons C + D \; \Delta H = -250\text{kj mol}^{-1}$$

250kj is absorbed when C and D react completely to give A and B.

This shows that 250 kJ is evolved (hence the negative sign) when 1 mole of A reacts completely with 2 moles of B. For reversible reactions, the value is always given as if the reaction was one-way in the forward direction.

The back reaction (the conversion of C and D into A and B) would be endothermic by exactly the same amount.

WHAT WOULD HAPPEN IF YOU CHANGED THE CONDITIONS BY INCREASING THE TEMPERATURE

According to Le Chatelier, the position of equilibrium will move in such a way as to counteract the change. That means that the position of equilibrium will move so that the temperature is reduced again.

Suppose the system is in equilibrium at 300°C, and you increase the temperature to 500°C. How can the reaction counteract the change you have made? How can it cool itself down again?

To cool down, it needs to absorb the extra heat that you have just put in. In the case we are looking at, the *back reaction* absorbs heat. The position of equilibrium therefore moves to the left. The new equilibrium mixture contains more A and B, and less C and D.

$$A + 2B \rightleftharpoons C + D \quad \Delta H = -250 \text{kj mol}^{-1}$$

$$\longleftarrow$$

The position of equilibrium moves to the left if you increase the temperature

If you were aiming to make as much C and D as possible, increasing the temperature on a reversible reaction where the forward reaction is exothermic isn't a good idea!

WHAT WOULD HAPPEN IF YOU CHANGED THE CONDITIONS BY DECREASING THE TEMPERATURE

The equilibrium will move in such a way that the temperature increases again.

Suppose the system is in equilibrium at 500°C and you reduce the temperature to 400°C. The reaction will tend to heat itself up again to return to the original temperature. It can do that by favouring the exothermic reaction.

The position of equilibrium will move to the right. More A and B are converted into C and D at the lower temperature.

$$A + 2B \rightleftharpoons C + D \quad \Delta H = -250 \text{kj mol}^{-1}$$

The position of equilibrium moves to the left if you decrease the temperature

- Increasing the temperature of a system in dynamic equilibrium favours the endothermic reaction. The system counteracts the change you have made by absorbing the extra heat.
- Decreasing the temperature of a system in dynamic equilibrium favours the exothermic reaction. The system counteracts the change you have made by producing more heat.

Again, this isn't in any way an *explanation* of why the position of equilibrium moves in the ways described. It is only a way of helping you to work out what happens

LE CHATELIER'S PRINCIPLE AND CATALYSTS

Catalysts have sneaked onto this page under false pretences, because *adding a catalyst makes absolutely no difference to the position of equilibrium,* and Le Chatelier's Principle doesn't apply to them.

This is because a catalyst speeds up the forward and back reaction to the same extent. Because adding a catalyst doesn't affect the relative rates of the two reactions, it can't affect the position of equilibrium. So why use a catalyst?

For a dynamic equilibrium to be set up, the rates of the forward reaction and the back reaction have to become equal. This doesn't happen instantly. For a very slow reaction, it could take years! A catalyst speeds up the rate at which a reaction reaches dynamic equilibrium.

Chapter 3

Redox Equilibria and Electrode Potentials

This page explains the background to standard electrode potentials (redox potentials), showing how they arise from simple equilibria, and how they are measured.

There are as many ways of teaching this as there are teachers and writers, and too many people make the fundamental mistake of forgetting that these are just simple equilibria. Too often, you will find the equations involved written as one-way rather than reversible. That small mistake makes the whole topic quite unnecessarily difficult to understand.

The approach you will find on this page (and whenever redox potentials are discussed on this site) avoids this problem completely by always talking in terms of equilibria.

THE DIFFERING REACTIVITIES OF METALS

When metals react, they give away electrons and form positive ions. This particular topic sets about comparing the ease with which a metal does this to form hydrated ions in solution - for example, $Mg^{2+}_{(aq)}$ or $Cu^{2+}_{(aq)}$.

We might want to compare the ease with which these two changes take place:

$$Mg_{(s)} \longrightarrow Mg^{2+}_{(aq)} + 2e^-$$

$$Cu_{(s)} \longrightarrow Cu^{2+}_{(aq)} + 2e^-$$

Everybody who has done chemistry for more than a few months knows that magnesium is more reactive than copper. The first reaction happens much more readily than the second one. What this topic does is to try to express this with some numbers.

LOOKING AT THIS FROM AN EQUILIBRIUM POINT OF VIEW

Suppose you have a piece of magnesium in a beaker of water. There will be some tendency for the magnesium atoms to shed electrons and go into solution as magnesium ions. The electrons will be left behind on the magnesium.

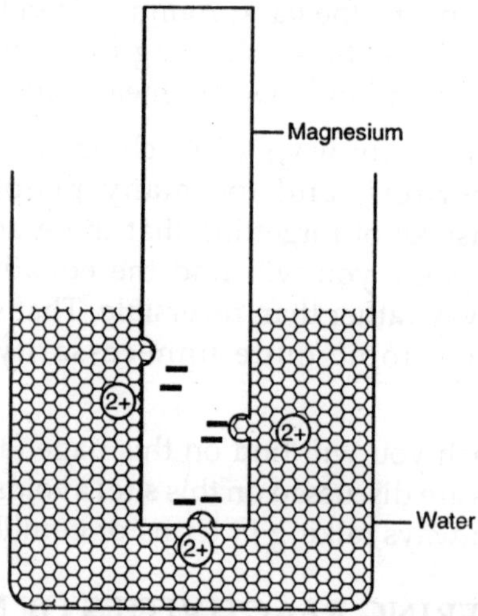

In a very short time, there will be a build-up of electrons on the magnesium, and it will be surrounded in the solution by a layer of positive ions. These will tend to stay close because they are attracted to the negative charge on the piece of metal.

Some of them will be attracted enough that they will reclaim their electrons and stick back on to the piece of metal.

A dynamic equilibrium will be established when the rate at which ions are leaving the surface is exactly equal to the rate at which they are joining it again. At that point there will be a constant negative charge on the magnesium, and a constant number of magnesium ions present in the solution around it.

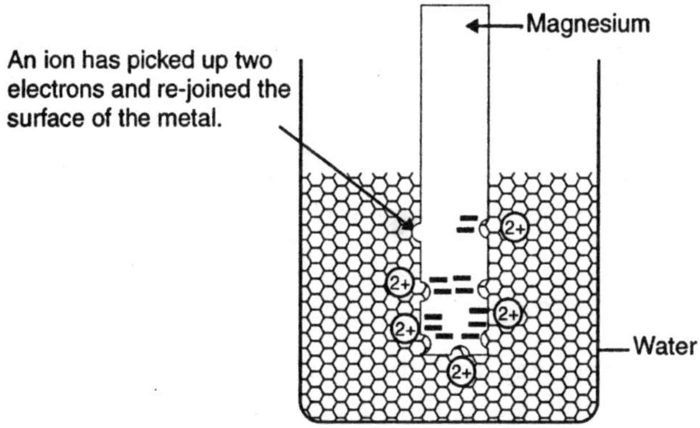

Simplifying the diagram to get rid of the "bites" out of the magnesium, you would be left with a situation like this:

Don't forget that this is just a snapshot of a dynamic equilibrium. Ions are continually leaving and rejoining the surface.

How would this be different if you used a piece of copper instead of a piece of magnesium?

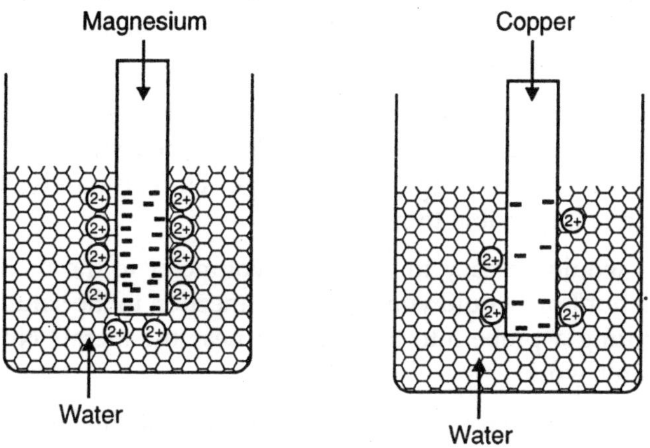

Copper is less reactive and so forms its ions less readily. Any ions which do break away are more likely to reclaim their electrons and stick back on to the metal again. You will still reach an equilibrium position, but there will be less charge on the metal, and fewer ions in solution.

If we write the two reactions as equilibria, then what we are doing is comparing the two positions of equilibrium.

The position of the magnesium equilibrium . . .

$$Mg^{2+}_{(aq)} + 2e^- \rightleftharpoons Mg_{(s)}$$

. . . lies further to the left than that of the copper equilibrium.

$$Cu^{2+}_{(aq)} + 2e^- \rightleftharpoons Cu_{(s)}$$

Notice the way that the two equilibria are written. By convention, all these equilibria are written with the electrons on the left-hand side of the equation. If you stick with this convention without fail, you will find that it makes the rest of this topic much easier to visualise.

Everything else concerning electrode potentials is simply an attempt to attach some numbers to these differing positions of equilibrium.

In principle, that is quite easy to do. In the magnesium case, there is a lot of difference between the negativeness of the metal and the positiveness of the solution around it. In the copper case, the difference is much less.

This *potential difference* could be recorded as a voltage - the bigger the difference between the positiveness and the negativeness, the bigger the voltage. Unfortunately, that voltage is impossible to measure!

It would be easy to connect a voltmeter to the piece of metal, but how would you make a connection to the solution? By putting a probe into the solution near the metal? No - it wouldn't work!

Any probe you put in is going to have a similar sort of equilibrium happening around it. The best you could measure would be some sort of combination of the effects at the probe and the piece of metal you are testing.

UNDERSTANDING THE IDEAS BEHIND A REFERENCE ELECTRODE

Suppose you had an optical device for measuring heights some distance away, and wanted to use it to find out how tall a particular person was. Unfortunately, you can't see their feet because they are standing in long grass.

Although you can't measure their absolute height, what you can do is to measure their height relative to the convenient post. Suppose that in this case, the person turned out to be 15 cm taller than the post.

You could repeat this for a range of people . . . . . . and come up with a set of results like this:

Person	Height relative to post (cm)
C	+20
A	+15
B	-15

Although you don't know any of their absolute heights, you can usefully rank them in order, and do some very simple sums to work out exactly how much taller one is than another. For example, C is 5 cm taller than A.

This turns out to be exactly what we need to do with the equilibria we started talking about. We don't actually need to know the absolute position of any of these equilibria. Going back to the magnesium and copper equilibria:

$$Mg^{2+}_{(aq)} + 2e^- \rightleftharpoons Mg_{(s)}$$

$$Cu^{2+}_{(aq)} + 2e^- \rightleftharpoons Cu_{(s)}$$

All we need to know is that the magnesium equilibrium lies further to the left than the copper one. We need to know that magnesium sheds electrons and forms ions more readily than copper does.

That means that we don't need to be able to measure the absolute voltage between the metal and the solution. It is enough to compare the voltage with a standardised system called a *reference electrode.*

The system used is called a standard hydrogen electrode.

MEASURING STANDARD ELECTRODE POTENTIALS (STANDARD REDOX POTENTIALS)

The Standard Hydrogen Electrode

The standard hydrogen electrode looks like this:

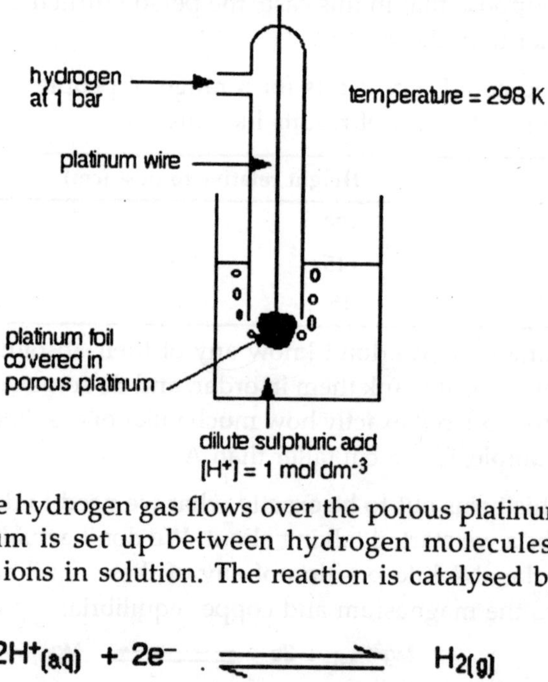

As the hydrogen gas flows over the porous platinum, an equilibrium is set up between hydrogen molecules and hydrogen ions in solution. The reaction is catalysed by the platinum.

$$2H^+_{(aq)} + 2e^- \rightleftharpoons H_{2(g)}$$

This is the equilibrium that we are going to compare all the others with.

The position of any equilibrium can be changed by changing conditions. That means that the conditions must be standardised so that you can make fair comparisons.

The hydrogen pressure is 1 bar (100 kPa). (You may find 1 atmosphere quoted in older sources.) The temperature is 298 K (25°C).

The concentration of the hydrogen ions in solution is also important. Changing concentrations is one of the ways of changing the position of an equilibrium. Throughout this topic, all ions concentrations are taken as being 1 mol dm^{-3}.

The standard hydrogen electrode is attached to the electrode system you are investigating - for example, a piece of magnesium in a solution containing magnesium ions.

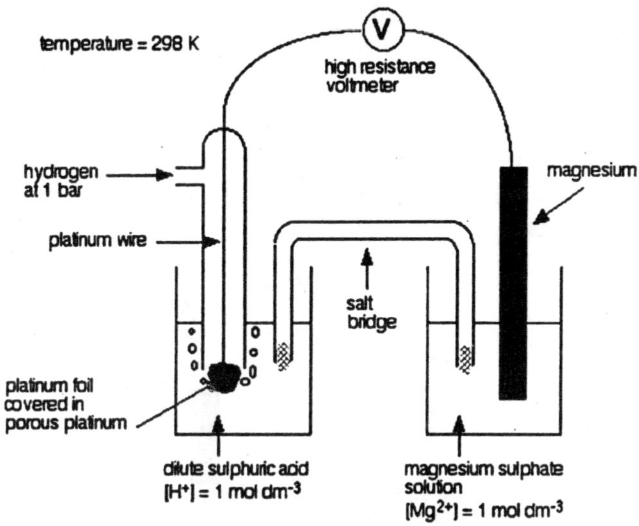

CELLS AND HALF CELLS

The whole of this set-up is described as a *cell*. It is a simple system which generates a voltage. Each of the two beakers and their contents are described as *half cells*.

The Salt Bridge

The salt bridge is included to complete the electrical circuit but without introducing any more bits of metal into the system. It is just a glass tube filled with an electrolyte like potassium nitrate solution. The ends are "stoppered" by bits of cotton wool. This stops too much mixing of the contents of the salt bridge with the contents of the two beakers.

The electrolyte in the salt bridge is chosen so that it doesn't react with the contents of either beaker.

These two equilibria are set up on the two electrodes (the magnesium and the porous platinum):

$$Mg^{2+}_{(aq)} + 2e^- \rightleftharpoons Mg_{(s)}$$

$$2H^+_{(aq)} + 2e^- \rightleftharpoons H_{2(g)}$$

Magnesium has a much greater tendency to form its ions than hydrogen does. The position of the magnesium equilibrium will be well to the left of that of the hydrogen equilibrium.

That means that there will be a much greater build-up of electrons on the piece of magnesium than on the platinum. Stripping all the rest of the diagram out, apart from the essential bits:

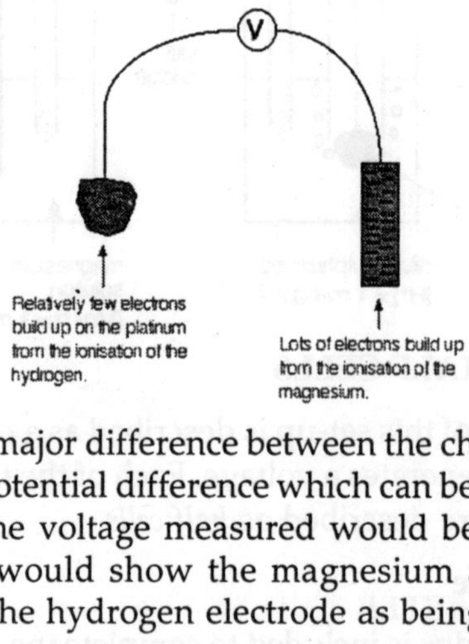

There is a major difference between the charge on the two electrodes - a potential difference which can be measured with a voltmeter. The voltage measured would be 2.37 volts and the voltmeter would show the magnesium as the negative electrode and the hydrogen electrode as being positive.

This sometimes confuses people! Obviously, the platinum in the hydrogen electrode isn't positive in real terms - there is a slight excess of electrons built up on it. But voltmeters don't deal in absolute terms - they simply measure a difference.

The magnesium has the greater amount of negativeness - the voltmeter records that as negative. The platinum of the hydrogen electrode isn't as negative - it is *relatively* more positive. The voltmeter records it as positive.

Throughout the whole of this redox potential work, you have to think in relative terms. For example, +0.4 is relatively more negative than +1.2. Or, another example, -0.3 is relatively more positive than -0.9.

This means replacing the magnesium half cell by one with a piece of copper suspended in a solution containing Cu^{2+} ions with a concentration of 1 mol dm^{-3}. You would probably choose to use copper(II) sulphate solution.

Copper forms its ions less readily than hydrogen does. Of the two equilibria . . .

$$Cu^{2+}_{(aq)} + 2e^- \rightleftharpoons Cu_{(s)}$$

$$2H^+_{(aq)} + 2e^- \rightleftharpoons H_{2(g)}$$

. . . the hydrogen one lies further to the left. That means that there will be less build-up of electrons on the copper than there is on the platinum of the hydrogen electrode.

There is less difference between the electrical charges on the two electrodes, so the voltage measured will be less. This time it is only 0.34 volts.

The other major change is that this time the copper is the more positive (less negative) electrode. The voltmeter will show the hydrogen electrode as the negative one and the copper electrode as positive.

THE VOLTMETER

You may have noticed that the voltmeter was described further up the page as having a "high resistance". Ideally, it wants to have an infinitely high resistance.

This is to avoid any flow of current through the circuit. If there was a low resistance in the circuit, electrons would flow from where there are a lot of them (around the magnesium, for example) to where there are less (on the hydrogen electrode).

If any current flows, the voltage measured drops. In order to make proper comparisons, it is important to measure the maximum possible voltage in any situation. This is called the *electromotive force* or *emf*.

The emf of a cell measured under standard conditions is given the symbol E°_{cell}.

CELL CONVENTIONS

Drawing a full diagram to represent a cell takes too long. Instead, the cell in which a magnesium electrode is coupled to a hydrogen electrode is represented like this:

Square brackets show the hydrogen flowing over the platinum.

Single vertical lines show a boundary between two phases - for example, between the magnesium ions and the solid metal.

$$Pt\ [H_{2(g)}]\ |\ 2H^+_{(aq)}\ ||\ Mg^{2+}_{(aq)}\ |\ Mg_{(s)}$$

Double vertical lines show the salt bridge. This is sometimes shown by a single (or double) *broken* vertical line (or lines).

You will often find variants on the way the hydrogen electrode is represented, such as:

$$Pt\ |\ H_{2(g)}\ |\ 2H^+_{(aq)}\ || \quad \text{or} \quad Pt\ [H_{2(g)}]\ |\ H^+_{(aq)}\ ||$$

Square brackets replaced by a vertical line showing the boundary between the platinum and the hydrogen.

One hydrogen ion shown rather than the 2 from the equation.

$$Pt\ [H_{2(g)}]\ |\ 2H^+_{(aq)}\ ||\ Mg^{2+}_{(aq)}\ |\ Mg_{(s)} \qquad E^o_{cell} = -2.37\ v$$

Shows that the magnesium (the right-hand electrode) is the negative one.

The convention is that you *show the sign of the right-hand electrode* (as you have drawn it) when you quote the E°_{cell} value. For example:

In the copper case:

$$Pt\,[H_{2(g)}]\ |\ 2H^+_{(aq)}\ ||\ Cu^{2+}_{(aq)}\ |\ Cu_{(s)} \qquad E^o_{cell} = +0.34\ v$$

Shows that the copper (i.e right-hand electrode) is the positive one.

Defining standard electrode potential (standard redox potential)

The values that we have just quoted for the two cells are actually the standard electrode potentials of the Mg^{2+} / Mg and Cu^{2+} / Cu systems.

The emf measured when a metal / metal ion electrode is coupled to a hydrogen electrode under standard conditions is known as the standard electrode potential of that metal / metal ion combination.

By convention, the hydrogen electrode is always written as the left-hand electrode of the cell. That means that the sign of the voltage quoted always gives you the sign of the metal electrode.

Standard electrode potential is given the symbol E°

Remember that the standard electrode potential of a metal / metal ion combination is the emf measured when that metal / metal ion electrode is coupled to a hydrogen electrode under standard conditions.

What you are doing is comparing the position of the metal / metal ion equilibrium with the equilibrium involving hydrogen.

Here are a few typical standard electrode potentials:

metal / metal ion combination	E° (volts)
Mg^{2+} / Mg	-2.37
Zn^{2+} / Zn	-0.76
Cu^{2+} / Cu	+0.34
Ag^+ / Ag	+0.80

Remember that each of these is comparing the position of the metal / metal ion equilibrium with the equilibrium involving hydrogen.

Here are the five equilibria (including the hydrogen one):

$$Mg^{2+}_{(aq)} + 2e^- \rightleftharpoons Mg_{(s)}$$

$$Zn^{2+}_{(aq)} + 2e^- \rightleftharpoons Zn_{(s)}$$

$$2H^+_{(aq)} + 2e^- \rightleftharpoons H_{2(g)}$$

$$Cu^{2+}_{(aq)} + 2e^- \rightleftharpoons Cu_{(s)}$$

$$Ag^+_{(aq)} + e^- \rightleftharpoons Ag_{(s)}$$

If you compare these with the E° values, you can see that the ones whose positions of equilibrium lie furthest to the left have the most negative E° values. That is because they form ions more readily - and leave more electrons behind on the metal, making it more negative.

Those which don't shed electrons as readily have positions of equilibrium further to the right. Their E° values get progressively more positive

REDOX POTENTIALS FOR NON-METAL AND OTHER SYSTEMS

This part explains how non-metals like chlorine can be included in the electrochemical series, and how other oxidising and reducing agents can have their standard electrode potentials (redox potentials) measured and fitted into the series.

MEASURING REDOX POTENTIALS FOR MORE COMPLICATED SYSTEMS

Systems Involving Gases

The obvious example here is chlorine. Chlorine is well known as an oxidising agent. Since the electrochemical series is about ranking substances according to their oxidising or reducing ability, it makes sense to include things like chlorine.

This time we are measuring the position of this equilibium relative to the hydrogen equilibrium.

$$Cl_{2(g)} + 2e^- \rightleftharpoons 2Cl^-_{(aq)}$$

Notice that the equilibrium is still written with the electrons on the left-hand side of the equation. That's why the chlorine gas has to appear on the left-hand side rather than on the right (which is where the metals and hydrogen appeared).

How can this equilibrium be connected into a circuit? The half-cell is built just the same as a hydrogen electrode. Chlorine gas is bubbled over a platinum electrode, which is immersed in a solution containing chloride ions with a concentration of 1 mol dm^{-3}.

The conventional way of writing the whole cell looks like this.

$$Pt_{(s)}\ [H_{2(g)}] \mid 2H^+_{(aq)} \parallel Cl_{2(g)},\ 2Cl^-_{(aq)} \mid Pt_{(s)}$$

$$E^o = +1.36\ v$$

Notice the way that the chlorine half cell is written. The convention is that the substance losing electrons is written closest to the electrode. In this case, the chloride ions are losing electrons.

If you had the chlorine half cell on the left-hand side in a different situation, then the convention still has to hold. The half cell would then be written:

$$Pt_{(s)} \mid 2Cl^-_{(aq)},\ Cl_{2(g)} \parallel \boxed{\text{other half cell}}$$

↑ chloride ions written dosest to the platinum electrode

↑ salt bridge

The value is positive and moderately high as E° values go. That means that the position of the Cl_2 / Cl^- equilibrium lies more to the right than the hydrogen equilibrium. Chlorine is much more likely to pick up electrons than hydrogen ions are.

$$2H^+_{(aq)} + 2e^- \rightleftharpoons H_{2(g)}$$

$$Cl_{2(g)} + 2e^- \rightleftharpoons 2Cl^-_{(aq)}$$

This equilibrium lies further to the right. Chlorine gas picks up electrons much more readily than hydrogen ions do.

Chlorine is therefore quite good a removing electrons from other things. It is a good oxidising agent.

MEASURING REDOX POTENTIALS FOR OTHER SYSTEMS

The Fe^{2+} / Fe^{3+} System

Iron(II) ions are easily oxidised to iron(III) ions, and iron(III) ions are fairly easily reduced to iron(II) ions. The equilibrium we are interested in this time is:

$$Fe^{3+}_{(aq)} + e^- \rightleftharpoons Fe^{2+}_{(aq)}$$

To measure the redox potential of this, you would simply insert a platinum electrode into a beaker containing a solution containing both iron(II) and iron(III) ions (1 mol dm^{-3} with respect to both), and couple this to a hydrogen electrode.

The cell diagram would look like this:

$$Pt_{(s)}\ [H_{2(g)}]\ |\ 2H^+_{(aq)}\ ||\ Fe^{3+}_{(aq)},\ Fe^{2+}_{(aq)}\ |\ Pt_{(s)}$$

$$E^\circ = +0.77\ v$$

Notice that the E° value isn't as positive as the chlorine one. The position of the iron(III) / iron(II) equilibrium isn't as far to the right as the chlorine equilibrium. That means that Fe^{3+} ions don't pick up electrons as easily as chlorine does. Chlorine is a stronger oxidising agent than Fe^{3+} ions.

POTASSIUM DICHROMATE(VI) AS AN OXIDISING AGENT

A commonly used oxidising agent, especially in organic chemistry, is potassium dichromate(VI) solution acidified with dilute sulphuric acid. The potassium ions are just spectator ions and aren't involved in the equilibrium in any way.

The equilibrium is more complicated this time because it contains more things:

$$Cr_2O_7^{2-}{}_{(aq)} + 14H^+{}_{(aq)} + 6e^- \rightleftharpoons 2Cr^{3+}{}_{(aq)} + 7H_2O_{(l)}$$

The half cell would have a piece of platinum dipping into a solution containing all the ions (dichromate(VI) ions, hydrogen ions and chromium(III) ions) all at 1 mol dm^{-3}.

There is yet another convention when it comes to writing these more complicated cell diagrams. Where there is more than one thing on either side of the equilibrium, square brackets are written around them to keep them tidy. The substances losing electrons are written next to the electrode, just as before.

$$\text{hydrogen electrode} \parallel [Cr_2O_7^{2-}{}_{(aq)} + 14H^+{}_{(aq)}],\ [2Cr^{3+}{}_{(aq)} + 7H_2O_{(l)}] \mid Pt_{(s)}$$

$$E^o = +1.33\ v$$

INCLUDING THESE NEW REDOX POTENTIALS IN THE ELECTROCHEMICAL SERIES

These values can be slotted seamlessly into the electrochemical series that so far has only included metals and hydrogen.

An updated electrochemical series

equilibrium			E° (volts)
$Li^+{}_{(aq)} + e^-$	$\rightleftharpoons$	$Li_{(s)}$	-3.03
$K^+{}_{(aq)} + e^-$	$\rightleftharpoons$	$K_{(s)}$	-2.92
$Ca^{2+}{}_{(aq)} + 2e^-$	$\rightleftharpoons$	$Ca_{(s)}$	-2.87

$Na^{+}_{(aq)} + e^{-}$	$\rightleftharpoons$	$Na_{(s)}$	-2.71
$Mg^{2+}_{(aq)} + 2e^{-}$	$\rightleftharpoons$	$Mg_{(s)}$	-2.37
$Al^{3+}_{(aq)} + 3e^{-}$	$\rightleftharpoons$	$Al_{(s)}$	-1.66
$Zn^{2+}_{(aq)} + 2e^{-}$	$\rightleftharpoons$	$Zn_{(s)}$	-0.76
$Fe^{2+}_{(aq)} + 2e^{-}$	$\rightleftharpoons$	$Fe_{(s)}$	-0.44
$Pb^{2+}_{(aq)} + 2e^{-}$	$\rightleftharpoons$	$Pb_{(s)}$	-0.13
$2H^{+}_{(aq)} + 2e^{-}$	$\rightleftharpoons$	$H_{2(g)}$	0
$Cu^{2+}_{(aq)} + 2e^{-}$	$\rightleftharpoons$	$Cu_{(s)}$	+0.34
$Fe^{3+}_{(aq)} + e^{-}$	$\rightleftharpoons$	$Fe^{2+}_{(aq)}$	+0.77
$Ag^{+}_{(aq)} + e^{-}$	$\rightleftharpoons$	$Ag_{(s)}$	+0.80
$Cr_2O_7^{2-}{}_{(aq)} + 14H^{+}_{(aq)} + 6e^{-}$	$\rightleftharpoons$	$2Cr^{3+}_{(aq)} + 7H_2O_{(l)}$	+1.33
$Cl_{2(g)} + 2e^{-}$	$\rightleftharpoons$	$2Cl^{-}_{(aq)}$	+1.36
$Au^{3+}_{(aq)} + 3e^{-}$	$\rightleftharpoons$	$Au_{(s)}$	+1.50

By coincidence, all the new equilibria we've looked at have positive E° values. It so happens that most of the equilibria with *negative* E° values that you meet at this level are ones involving simple metal / metal ion combinations.

REDOX POTENTIALS AND SIMPLE TEST TUBE REACTIONS

This page explains how standard electrode potentials (redox potentials) relate to simple and familiar reactions that can be done in test tubes.

COMBINING A ZINC WITH A COPPER HALF CELL

So far in this series of pages, we have looked at combinations of a hydrogen electrode with the half cell we have been interested in. However, there isn't any reason why you can't couple *any two* half cells together.

This next bit looks at what happens if you combine a zinc half cell with a copper half cell.

IN THE PRESENCE OF A HIGH RESISTANCE VOLTMETER

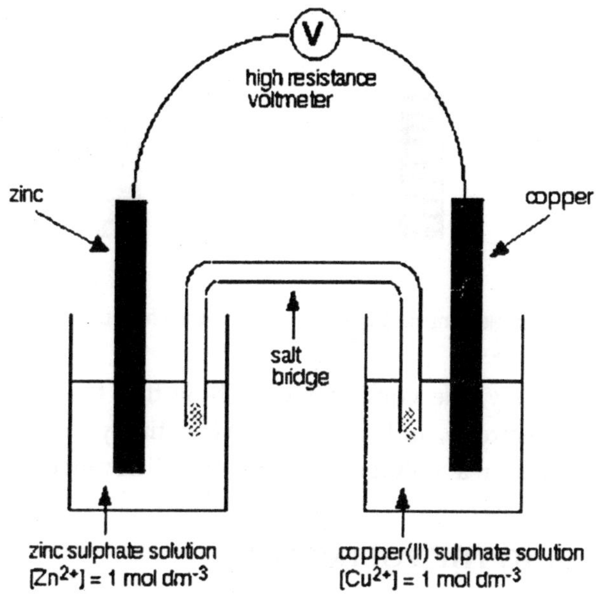

The two equilibria which are set up in the half cells are:

$$Zn^{2+}_{(aq)} + 2e^- \rightleftharpoons Zn_{(s)} \qquad E° = -0.76\ v$$

$$Cu^{2+}_{(aq)} + 2e^- \rightleftharpoons Cu_{(s)} \qquad E° = +0.34\ v$$

The negative sign of the zinc E° value shows that it releases electrons more readily than hydrogen does. The positive sign of the copper E° shows that it releases electrons less readily than hydrogen.

That means that you can compare any two equilibria directly. For example, in this case you can see that the zinc releases electrons more readily than the copper does - the position of the zinc equilibrium lies further to the left than the copper equilibrium.

Stripping everything else out of the diagram, and looking only at the build up of electrons on the two pieces of metal:

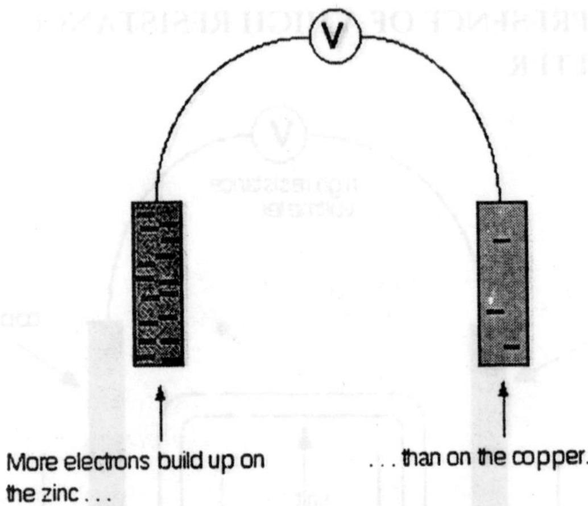

Obviously, the voltmeter will show that the zinc is the negative electrode, and copper is the (relatively) positive one. It will register a voltage showing the difference between them.

REMOVING THE VOLTMETER

The high resistance of the voltmeter is deliberately designed to stop any current flow in the circuit. What happens if you remove the voltmeter and replace it with a bit of wire?

Electrons will flow from where there are a lot of them (on the zinc) to where there are fewer (on the copper). The movement of the electrons is an electrical current.

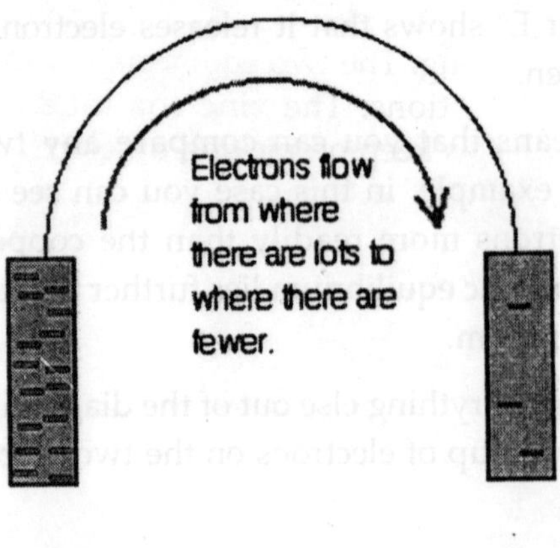

THE EFFECT OF THIS ON THE EQUILIBRIA

These are just simple equilibria, and you can apply Le Chatelier's Principle to them.

Electrons are flowing away from the zinc equilibrium. According to Le Chatelier's Principle, the position of equilibrium will move to replace the lost electrons.

Electrons are being dumped onto the piece of copper in the copper equilibrium. According to Le Chatelier's Principle, the position of equilibrium will move to remove these extra electrons.

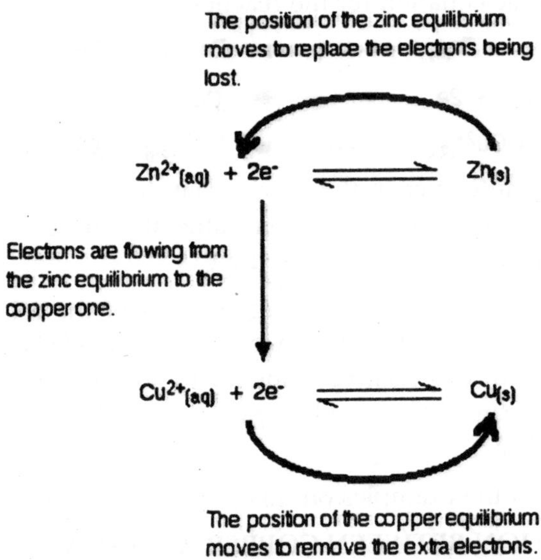

If electrons continue to flow, the positions of equilibrium keep on shifting. The two equilibria essentially turn into two one-way reactions. The zinc continues to ionise, and the copper(II) ions keep on picking up electrons.

$$Zn_{(s)} \longrightarrow Zn^{2+}_{(aq)} + 2e^-$$

$$Cu^{2+}_{(aq)} + 2e^- \longrightarrow Cu_{(s)}$$

Taking the apparatus as a whole, there is a chemical reaction going on in which zinc is going into solution as zinc ions, and is giving electrons to copper(II) ions to turn them into metallic copper.

RELATING THIS TO A TEST TUBE REACTION

This is exactly the same reaction that occurs when you drop a piece of zinc into some copper(II) sulphate solution. The blue colour of the solution fades as the copper(II) ions are converted into brown copper metal. The final solution contains zinc sulphate. (The sulphate ions are spectator ions.)

You can add the two electron-half-equations above to give the overall ionic equation for the reaction.

$$Zn_{(s)} \longrightarrow Zn^{2+}_{(aq)} + 2e^-$$

$$Cu^{2+}_{(aq)} + 2e^- \longrightarrow Cu_{(s)}$$

$$Zn_{(s)} + Cu^{2+}_{(aq)} \longrightarrow Zn^{2+}_{(aq)} + Cu_{(s)}$$

The only difference in this case is that the zinc gives the electrons *directly* to the copper(II) ions rather than the electrons having to travel along a bit of wire first.

The test tube reaction happens because of the relative tendency of the zinc and copper to lose electrons to form ions. You can find out this relative tendency by looking at the E° values. That means that any redox reaction could be discussed in a similar way.

The rest of the examples on this page illustrate this.

THE REACTION BETWEEN COPPER AND SILVER NITRATE SOLUTION

If you hang a coil of copper wire in some colourless silver nitrate solution, the copper gets covered in silver - partly as a grey fur, and partly as delicate crystals. The solution turns blue.

Building the ionic equation from the two half-equations:

$$Cu_{(s)} \longrightarrow Cu^{2+}_{(aq)} + 2e^-$$

$$2 \times (\ Ag^+_{(aq)} + e^- \longrightarrow Ag_{(s)}\ )$$

$$Cu_{(s)} + 2Ag^+_{(aq)} \longrightarrow Cu^{2+}_{(aq)} + 2Ag_{(s)}$$

WORKING FROM THE REDOX POTENTIALS

How does this relate to the E° values for copper and silver?

$$Cu^{2+}_{(aq)} + 2e^- \rightleftharpoons Cu_{(s)} \qquad E^\circ = +0.34\ v$$

$$Ag^{+}_{(aq)} + e^- \rightleftharpoons Ag_{(s)} \qquad E^\circ = +0.80\ v$$

You can see that both of these E° values are positive. Neither copper nor silver produce ions and release electrons as easily as hydrogen does.

However, of the two, copper releases electrons more readily. In a cell, the copper would have the greater build up of electrons, and be the negative electrode. If the copper and zinc were connected by a bit of wire, electrons would flow from the copper to the silver.

That, of course, will upset the two equilibria:

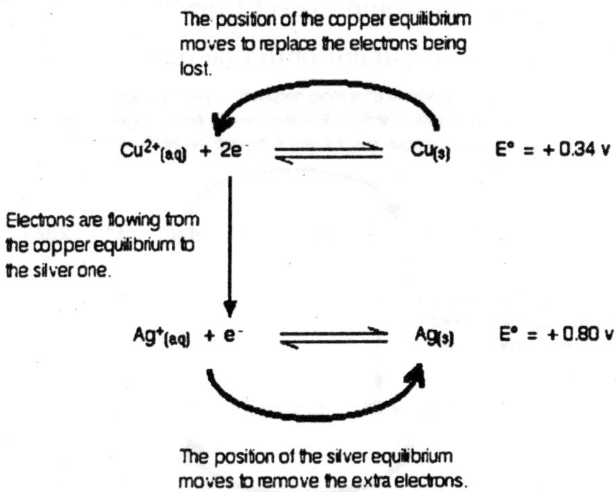

Electrons will continue to flow and the two equilibria will again turn into one-way reactions to give the electron-half-equations we've just used to build the ionic equation. Showing that again:

$$Cu_{(s)} \longrightarrow Cu^{2+}_{(aq)} + 2e^-$$

$$2 \times (\ Ag^{+}_{(aq)} + e^- \longrightarrow Ag_{(s)}\ )$$

$$Cu_{(s)} + 2Ag^{+}_{(aq)} \longrightarrow Cu^{2+}_{(aq)} + 2Ag_{(s)}$$

Whenever you link two of these equilibria together (either via a bit of wire, or by allowing one of the substances to give electrons directly to another one in a test tube):

- The equilibrium with the more negative (or less positive) E° value will move to the left.
- The equilibrium with the more positive (or less negative) E° value will move to the right.

It sounds simple, but that is the key to using E° values in most of the situations you will come across.

THE REACTION BETWEEN MAGNESIUM AND DILUTE SULPHURIC ACID

Magnesium reacts with dilute sulphuric acid to give hydrogen and a colourless solution containing magnesium sulphate.

Is this what you would expect from the E° values?

Putting all the argument onto one diagram:

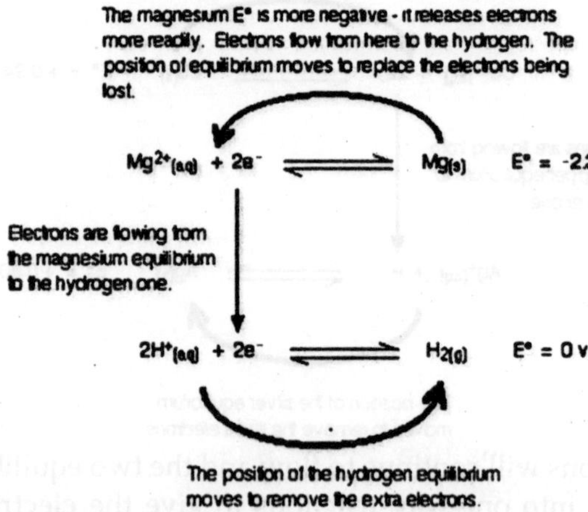

The two equilibria become one-way reactions which you can use to build the ionic equation:

$$Mg_{(s)} \longrightarrow Mg^{2+}_{(aq)} + 2e^-$$

$$2H^+_{(aq)} + 2e^- \longrightarrow H_{2(g)}$$

$$Mg_{(s)} + 2H^+_{(aq)} \longrightarrow Mg^{2+}_{(aq)} + H_{2(g)}$$

USING POTASSIUM DICHROMATE(VI) AS AN OXIDISING AGENT

Potassium dichromate(VI) acidified with dilute sulphuric acid oxidises iron(II) ions to iron(III) ions. The orange solution containing the dichromate(VI) ions turns green as chromium(III) ions are formed.

How does this relate to the E° values?

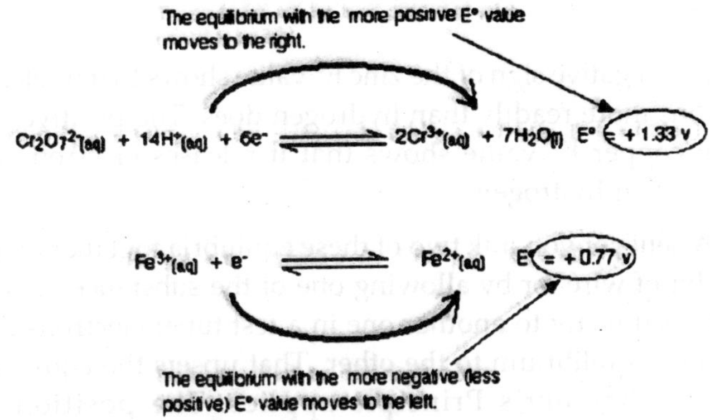

Building the ionic equation then works like this:

$$Cr_2O_7^{2-}{}_{(aq)} + 14H^+{}_{(aq)} + 6e^- \longrightarrow 2Cr^{3+}{}_{(aq)} + 7H_2O_{(l)}$$

$$6 \times (\ Fe^{2+}{}_{(aq)} \longrightarrow Fe^{3+}{}_{(aq)} + e^-\ )$$

$$Cr_2O_7^{2-}{}_{(aq)} + 14H^+{}_{(aq)} + 6Fe^{2+}{}_{(aq)} \longrightarrow 2Cr^{3+}{}_{(aq)} + 7H_2O_{(l)} + 6Fe^{3+}{}_{(aq)}$$

MAKING PREDICTIONS USING REDOX POTENTIALS (ELECTRODE POTENTIALS)

This explains how to use redox potentials (electrode potentials) to predict the feasibility of redox reactions. It also looks at how you go about choosing a suitable oxidising agent or reducing agent for a particular reaction.

PREDICTING THE FEASIBILITY OF A POSSIBLE REDOX REACTION

Standard electrode potentials (redox potentials) are one way of measuring how easily a substance loses electrons. In particular, they give a measure of relative positions of equilibrium in reactions such as:

$$Zn^{2+}_{(aq)} + 2e^- \rightleftharpoons Zn_{(s)} \qquad E^\circ = -0.76\ v$$

$$Cu^{2+}_{(aq)} + 2e^- \rightleftharpoons Cu_{(s)} \qquad E^\circ = +0.34\ v$$

The more negative the E° value, the further the position of equilibrium lies to the left. Remember that this is always relative to the hydrogen equilibrium - and not in absolute terms.

The negative sign of the zinc E° value shows that it releases electrons more readily than hydrogen does. The positive sign of the copper E° value shows that it releases electrons less readily than hydrogen.

Whenever you link two of these equilibria together (either via a bit of wire, or by allowing one of the substances to give electrons directly to another one in a test tube) electrons flow from one equilibrium to the other. That upsets the equilibria, and Le Chatelier's Principle applies. The positions of equilibrium move - and keep on moving if the electrons continue to be transferred.

The two equilibria essentially turn into two one-way reactions:

- The equilibrium with the more negative (or less positive) E° value will move to the left.
- The equilibrium with the more positive (or less negative) E° value will move to the right.

Note: If you aren't confident about this, *please* go back and start from the beginning of this sequence of pages. All these ideas are explored in a gentle and logical way. Links at the bottom of each page will bring you back here again.

Of course it does! I'm choosing this as an introductory example because everybody will know the right answer before we start. We have also explored this from a slightly different point of view on the previous page in this sequence.

The E° values are:

$$Mg^{2+}_{(aq)} + 2e^- \rightleftharpoons Mg_{(s)} \qquad E° = -2.37\ v$$

$$2H^+_{(aq)} + 2e^- \rightleftharpoons H_{2(g)} \qquad E° = 0\ v$$

You are starting with magnesium metal and hydrogen ions in the acid. The sulphate ions are spectator ions and play no part in the reaction.

Think of it like this. There is no need to write anything down unless you want to. With a small amount of practice, all you need to do is just look at the numbers.

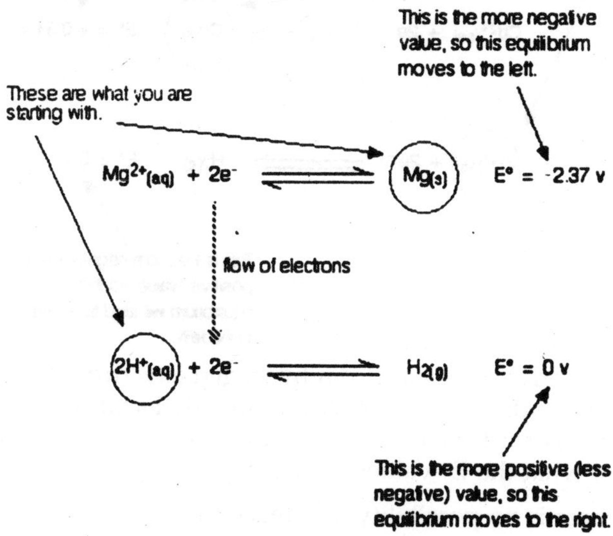

Is there anything to stop the sort of movements we have suggested? No! The magnesium can freely turn into magnesium ions and give electrons to the hydrogen ions producing hydrogen gas. The reaction is feasible.

Now for a reaction which turns out not to be feasible . . .

You know that the answer is that it won't. How do the E° values predict this?

$$Cu^{2+}_{(aq)} + 2e^- \rightleftharpoons Cu_{(s)} \qquad E° = +0.34\ v$$

$$2H^+_{(aq)} + 2e^- \rightleftharpoons H_{2(g)} \qquad E° = 0\ v$$

Doing the same sort of thinking as before:

The diagram shows the way that the E° values are telling us that the equilibria will tend to move. Is this possible? No!

If we start from copper metal, the copper equilibrium is *already* completely to the right. If it were to react at all, the equilibrium will have to move to the left - directly opposite to what the E° values are saying.

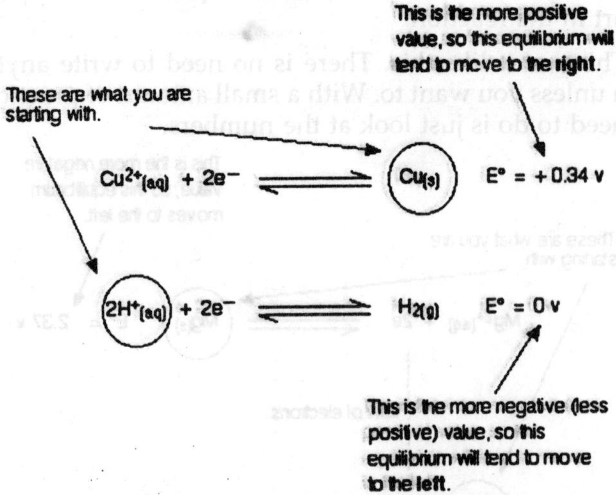

Similarly, if we start from hydrogen ions (from the dilute acid), the hydrogen equilibrium is already as far to the left as possible. For it to react, it would have to move to the right - against what the E° values demand.

There is no possibility of a reaction.

In the next couple of examples, decide for yourself whether or not the reaction is feasible *before* you read the text.

WILL OXYGEN OXIDISE IRON(II) HYDROXIDE TO IRON(III) HYDROXIDE UNDER ALKALINE CONDITIONS

The E° values are:

$$Fe(OH)_{3(s)} + e^- \rightleftharpoons Fe(OH)_{2(s)} + OH^-_{(aq)} \qquad E° = -0.56\ v$$

$$O_{2(g)} + 2H_2O_{(l)} + 4e^- \rightleftharpoons 4OH^-_{(aq)} \qquad E° = +0.40\ v$$

Think about this before you read on. Remember that the equilibrium with the more negative E° value will tend to move to the left. The other one tends to move to the right. Is that possible?

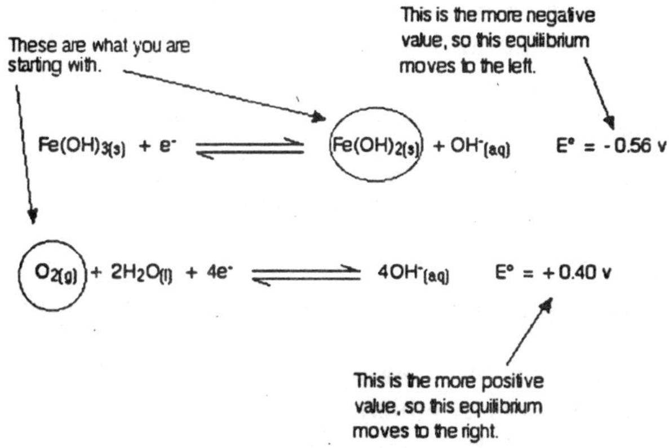

Yes, it is possible. Given what we are starting with, both of these equilibria *can* move in the directions required by the E° values. The reaction is feasible.

CHLORINE OXIDISE MANGANESE(II) IONS TO MANGANATE(VII) IONS

The E° values are:

$$MnO_4^-{}_{(aq)} + 8H^+{}_{(aq)} + 5e^- \rightleftharpoons Mn^{2+}{}_{(aq)} + 4H_2O_{(l)} \quad E° = +1.51\ v$$

$$Cl_{2(g)} + 2e^- \rightleftharpoons 2Cl^-{}_{(aq)} \quad E° = +1.36\ v$$

Again, think about this before you read on.

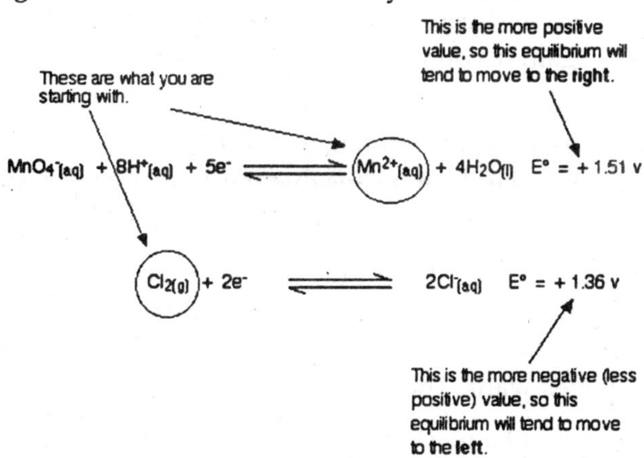

Given what you are starting from, these equilibrium shifts are impossible. The manganese equilibrium has the more positive E° value and so will tend to move to the right. However, because we are starting from manganese(II) ions, it is already as far to the right as possible. In order to get any reaction, the equilibrium would have to move to the *left*. That is against what the E° values are saying.

This reaction isn't feasible.

WILL DILUTE NITRIC ACID REACT WITH COPPER

This is going to be more complicated because there are two different ways in which dilute nitric acid might possibly react with copper. The copper might react with the hydrogen ions or with the nitrate ions. Nitric acid reactions are always more complex than the simpler acids like sulphuric or hydrochloric acid because of this problem.

Here are the E° values:

$$Cu^{2+}_{(aq)} + 2e^- \rightleftharpoons Cu_{(s)} \qquad E^\circ = +0.34\ v$$

$$2H^+_{(aq)} + 2e^- \rightleftharpoons H_{2(g)} \qquad E^\circ = 0\ v$$

$$NO^-_{3(aq)} + 4H^+_{(aq)} + 3e^- \rightleftharpoons NO_{(g)} + 2H_2O_{(l)} \qquad E^\circ = +0.96\ v$$

We have already discussed the possibility of copper reacting with hydrogen ions further up this page. Go back and look at it again if you need to, but the argument (briefy) goes like this:

The copper equilibrium has a more positive E° value than the hydrogen one. That means that the copper equilibium will tend to move to the right and the hydrogen one to the left.

However, if we start from copper and hydrogen ions, the equilibria are already as far that way as possible. Any reaction would need them to move in the opposite direction to what the E° values want. The reaction isn't feasible.

What about a reaction between the copper and the nitrate ions?

This is feasible. The nitrate ion equilibrium has the more positive E° value and will move to the right. The copper E° value is less positive. That equilibrium will move to the left. The movements that the E° values suggest are possible, and so a reaction is feasible.

Copper(II) ions are produced together with nitrogen monoxide gas.

It sometimes happens that E° values suggest that a reaction ought to happen, but it doesn't. Occasionally, a reaction happens although the E° values seem to be the wrong way around. These next two examples explain how that can happen. By coincidence, both involve the dichromate(VI) ion in potassium dichromate(VI).

ACIDIFIED POTASSIUM DICHROMATE(VI) OXIDISE WATER

The E° values are:

$$Cr_2O_7^{2-}{}_{(aq)} + 14H^+{}_{(aq)} + 6e^- \rightleftharpoons 2Cr^{3+}{}_{(aq)} + 7H_2O_{(l)} \quad E° = +1.33\ v$$

$$O_{2(g)} + 4H^+{}_{(aq)} + 4e^- \rightleftharpoons 2H_2O_{(l)} \quad E° = +1.23\ v$$

The relative sizes of the E° values show that the reaction is feasible:

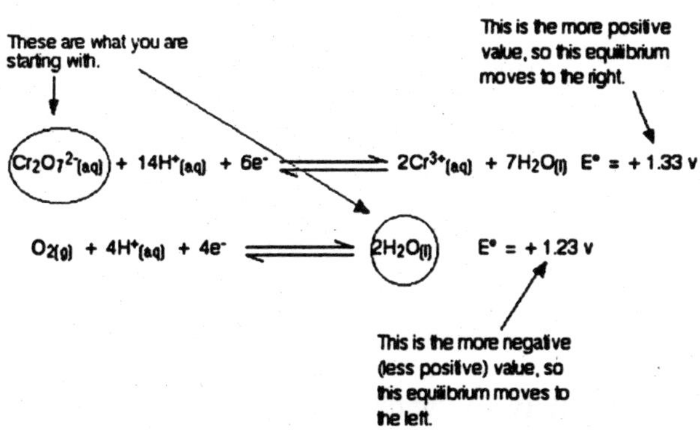

However, in the test tube nothing happens however long you wait. An acidified solution of potassium dichromate(VI) doesn't react with the water that it is dissolved in. So what is wrong with the argument?

In fact, there is nothing wrong with the argument. The problem is that all the E° values show is that a reaction is *possible.* They don't tell you that it will actually happen. There may be very large activation barriers to the reaction which prevent it from taking place.

Always treat what E° values tell you with some caution. All they tell you is whether a reaction is *feasible* - they tell you nothing about how fast the reaction will happen.

ACIDIFIED POTASSIUM DICHROMATE(VI) OXIDISE CHLORIDE IONS TO CHLORINE

The E° values are:

$$Cr_2O_7^{2-}{}_{(aq)} + 14H^+{}_{(aq)} + 6e^- \rightleftharpoons 2Cr^{3+}{}_{(aq)} + 7H_2O_{(l)} \quad E^\circ = +1.33\ v$$

$$Cl_{2(g)} + 2e^- \rightleftharpoons 2Cl^-{}_{(aq)} \quad E^\circ = +1.36\ v$$

Because the chlorine E° value is slightly greater than the dichromate(VI) one, there shouldn't be any reaction. For a reaction to occur, the equilibria would have to move in the wrong directions.

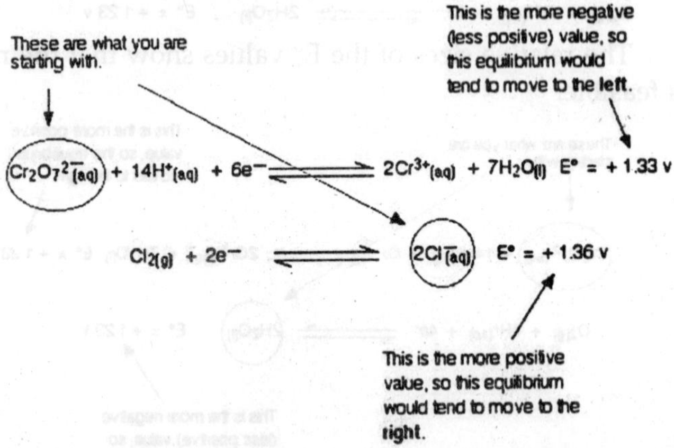

Unfortunately, in the test tube, potassium dichromate(VI) solution *does* oxidise concentrated hydrochloric acid to chlorine. The hydrochloric acid serves as the source of the hydrogen ions in the dichromate(VI) equilibrium and of the chloride ions.

The problem here is that E° values only apply under *standard* conditions. If you change the conditions you will change the position of an equilibrium - and that will change its E value. (Notice that you can't call it an E° value any more, because the conditions are no longer standard.)

The standard condition for concentration is 1 mol dm^{-3}. But concentrated hydrochloric acid is approximately 10 mol dm^{-3}. The concentrations of the hydrogen ions and chloride ions are far in excess of standard.

What effect does that have on the two positions of equilibrium?

$$Cr_2O_7^{2-}{}_{(aq)} + 14H^+_{(aq)} + 6e \rightleftharpoons 2Cr^{3+}_{(aq)} + 7H_2O_{(l)} \quad E° = +1.33\ v$$

Increasing the concentration of these will move the position of equilibrium to the right (Le Chatelier's Principle).

That will make the E value even more positive.

$$Cl_{2(g)} + 2e^- \rightleftharpoons 2Cl^-_{(aq)} \quad E° = +1.36\ v$$

Increasing the concentration of these will move the position of equilibrium to the left (Le Chatelier's Principle).

That will make the E value less positive.

Because the E° values are so similar, you don't have to change them very much to make the dichromate(VI) one the more positive. As soon as that happens, it will react with the chloride ions to produce chlorine.

In most cases, there is enough difference between E° values that you can ignore the fact that you aren't doing a reaction under strictly standard conditions. But sometimes it does make a difference. Be careful!

Selecting an oxidising or reducing agent for a reaction

There is nothing remotely new in this. It is just a slight variation on what we've just been looking at.

Choosing an Oxidising Agent

Remember:

- Oxidation is loss of electrons.
- An oxidising agent oxidises something by removing electrons from it. That means that the oxidising agent gains electrons.

It is easier to explain this with a specific example. What could you use to oxidise iron(II) ions to iron(III) ions? The E° value for this reaction is:

$$Fe^{3+}_{(aq)} + e^- \rightleftharpoons Fe^{2+}_{(aq)} \quad E° = +0.77\ v$$

To change iron(II) ions into iron(III) ions, you need to persuade this equilibrium to move to the left. That means that when you couple it to a second equilibrium, this iron E° value must be the more negative (less positive one).

You could use anything which has a more positive E° value. For example, you could use any of the these which we have already looked at over the last page or two:

Dilute nitric acid:

$$NO_3^-{}_{(aq)} + 4H^+_{(aq)} + 3e^- \rightleftharpoons NO_{(g)} + 2H_2O_{(l)} \quad E° = +0.96\ v$$

Acidified potassium dichromate(VI):

$$Cr_2O_7^{2-}{}_{(aq)} + 14H^+_{(aq)} + 6e^- \rightleftharpoons 2Cr^{3+}_{(aq)} + 7H_2O_{(l)} \quad E° = +1.33\ v$$

Chlorine:

$$Cl_{2(g)} + 2e^- \rightleftharpoons 2Cl^-_{(aq)} \quad E° = +1.36\ v$$

Acidified potassium manganate(VII):

$$MnO_4^-{}_{(aq)} + 8H^+_{(aq)} + 5e^- \rightleftharpoons Mn^{2+}_{(aq)} + 4H_2O_{(l)} \quad E° = +1.51\ v$$

Choosing a Reducing Agent

Remember:

- Reduction is gain of electrons.
- A reducing agent reduces something by giving electrons to it. That means that the reducing agent loses electrons.

You have to be a little bit more careful this time, because the substance losing electrons is found on the right-hand side of one of these redox equilibria. Again, a specific example makes it clearer.

For example, what could you use to reduce chromium(III) ions to chromium(II) ions? The E° value is:

$$Cr^{3+}_{(aq)} + e^- \rightleftharpoons Cr^{2+}_{(aq)} \quad E^\circ = -0.41\ v$$

You need this equilibrium to move to the right. That means that when you couple it with a second equilibrium, this chromium E° value must be the most positive (least negative).

In principle, you could choose anything with a *more negative* E° value - for example, zinc:

$$Zn^{2+}_{(aq)} + 2e^- \rightleftharpoons Zn_{(s)} \quad E^\circ = -0.76\ v$$

You would have to remember to start from metallic zinc, and not zinc ions. You need this second equilibrium to be able to move to the left to provide the electrons. If you started with zinc ions, it would already be on the left - and would have no electrons to give away. Nothing could possibly happen if you mixed chromium(III) ions and zinc ions.

That is fairly obvious in this simple case. If you were dealing with a more complicated equilibrium, you would have to be careful to think it through properly.

THE VIRIAL EXPANSION

The virial expansion, also called the virial equation of state, is the most interesting and versatile of the equations of state for gases.. The virial expansion is a power series in powers of the variable, *n*/*V*, and has the form,

$$\frac{pV}{nRT} = Z = 1 + B(T)\frac{n}{V} + C(T)\frac{n^2}{V^2} + \ldots \infty \tag{1}$$

The coefficient, *B*(*T*), is a function of temperature and is called the "second virial coefficient. *C*(*T*) is called the third virial coefficient, and so on. The expansion is, in principle, an infinite series, and as such should be valid for all isotropic substances. In practice, however, terms above the third virial coefficient are rarely used in chemical thermodynamics.

Notice that we have set the quantity pV/nRT equal to Z. This quantity (Z) is called the "compression factor." It is a useful measure of the deviation of a real gas from an ideal gas. For an ideal gas the compression factor is equal to 1.

THE BOYLE TEMPERATURE

The second virial coefficient, *B(T)*, is an increasing function of temperature throughout most of the useful temperature range. (It does decrease slightly at very high temperatures.) *B* is negative at low temperatures, passes through zero at the so-called "Boyle temperature," and then becomes positive. The temperature at which $B(T) = 0$ is called the Boyle temperature because the gas obeys Boyle's law to high accuracy at this temperature. We can see this by noting that at the Boyle temperature the virial expansion looks like,

$$\frac{pV}{nRT} = 1 + 0 + C\frac{n^2}{V^2} + \ldots\infty \tag{2}$$

If the density is not too high the C term is very small so that the system obeys Boyle's law.

ALTERNATE FORM OF THE VIRIAL EXPANSION

An equivalent form of the virial expansion is an infinite series in powers of the pressure.

$$\frac{pV}{nRT} = 1 + B'(T)p + C'(T)p^2 + \ldots\infty \tag{3}$$

The new virial coefficients, *B′*, *C′*, . . ., can be calculated from the original virial coeffients, *B*, *C*, ... To do this we equate the two virial expansions,

$$\frac{pV}{nRT} = 1 + B(T)\frac{n}{V} + C(T)\frac{n^2}{V^2} + \ldots\infty$$

$$= 1 + B'(T)p + C'(T)p^2 + \ldots\infty \tag{4}$$

Then we solve the original virial expansion for p,

$$p=\frac{nRT}{V}=(1+B(T)\frac{n}{V}+C(T)\frac{n^2}{V^2}+...\infty \tag{5}$$

and substitute this expression for *p* into the right-hand-side of equation (4),

$$1+B\frac{n}{V}+C\frac{n^2}{V^2}+...=1+B'\frac{nRT}{V}(1+B'\frac{n}{V}+C\frac{n^2}{V^2}+...)$$
$$+C'\left(\frac{nRT}{V}\right)^2\left(1+B\frac{n}{V}+C\frac{n^2}{V^2}+...\right)^2+... \tag{6a}$$

$$1+B\frac{n}{V}+C\frac{n^2}{V^2}+...=1+B'RT\frac{n}{V}+B'RT\frac{n^2}{V^2}+$$
$$...C'(RT)^2\frac{n^2}{V^2}+... \tag{6b}$$

Both sides of Equation (6b) are power series in *n*/*V*. (We have omitted third and higher powers of *n*/*V*because the second power is as high as we are going here.) Since the two power series must be equal, the coefficients of each power of *n*/*V* must be the same on both sides. The coefficient of $(n/V)^0$ on each side is 1, which gives the reassuring but not very interesting result, 1 = 1. Equating the coefficient of $(n/V)^1$ on each side gives $B = B'RT$ and equating the coefficients of $(n/V)^2$ gives

$$C=B'RTB+C'(RT)^2 \tag{7}$$

These equations are easily solved to give *B′* and *C′* in terms of *B*, *C*, and *R*.

$$B'=\frac{B}{RT}$$

$$C'=\frac{C-B^2}{(RT)^2} \tag{8}$$

Useful exercises would be:

1. Extend the two virial expansions to the D and D' terms respectively and find the expression for D' in terms of B, C, and D.
2. Find B' and C' in terms of the van der Waals a and b constants. (You were asked, in the homework to find the virial coefficients B and C in terms of a and b so you already have these.)

The word "virial" is related to the Latin word for force. Clausius (whose name we will see frequently) named a certain function of the force between molecules "the virial of force." This name was subsequently taken over for the virial expansion because the terms in that expansion can be calculated from the forces between the molecules.

The virial expansion is important for several reasons, among them: It can, in principle, be made as accurate as desired by keeping more terms. Also, it has a sound theoretical basis. The virial coefficients can be calculated from a theoretical model of the intermolecular potential energy of the gas molecules

CRITICAL PHENOMENA

All real gases can be liquefied. Depending on the gas this might require compression and/or cooling. However, there exists for each gas a temperature above which the gas cannot be liquefied. This temperature, above which the gas cannot be liquefied, is called the critical temperature and it is usually symbolized by, T_C . In order to liquefy a real gas the temperature must be at, or below, its critical temperature.

There are gases, sometimes called the "permanent gases" which have critical temperatures below room temperature. These gases must be cooled to a temperature below their critical point, which means below room temperture, before they can be liquefied. Examples of "permanent gases" include, He, H_2, N_2, O_2, Ne, Ar, and so on. Many substances have critical temperatures above room temperature. These substances exist as liquids (or even solids) at room temperature. Water, for example, has a critical temperature of 647.1 K, much higher than the 298.15 K standard room temperature. Water can be liquefied at any temperature below 647.1 K (although above 398.15 - the normal boiling point of water - you would have to apply a pressure higher than atmospheric temperature in order to keep it liquid.

It is convenient to think about liquefying substances and critical phenomena using a *p-V* diagram. This is a graph with pressure, *p*, plotted on the vertical axis and the volume, *V*, plotted along the horizontal axis. If we plot the pressure of a substance as a function of volume, holding temperature constant we get a series of curves, called isotherms. There is an example of such a plot in most physical chemistry texts.

Notice that when a substance is liquefied the isotherm becomes "flat," that is, the slope becomes zero. On the critical isotherm the slope "just barely" becomes flat at one point on the graph. A point where a decreasing function becomes flat before continuing to decrease is called a point of inflection. The mathematical characteristic of an inflection point is that the first and second derivatives are zero at that point. For our critical isotherm on a p-V diagram we would write,

$$\left(\frac{\partial p}{\partial V}\right)_T = 0 \quad (1)$$

and

$$\left(\frac{\partial^2 p}{\partial V^2}\right)_T = 0 \quad (2)$$

Equations (1) and (2) constitute a set of two equation in two unknowns, *V*, and *T*. One can test to see if an approximate equation of state gives a critical point by calculating these two derivatives for the equation of state and trying to solve the pair of equations. If a solution exists (and p and V are neither zero or infinity) then we say that the equation of state has a critical point.

Let's use this test to see if the ideal gas has a critical point. First we have to solve the ideal gas equation of state. $PV = nRT$, for pressure, p.

$$P = \frac{nRT}{V} \tag{3}$$

Now we can take the derivatives in Equations 1 and 2 and set them (independently[2]) equal to zero.

$$\left(\frac{\partial p}{\partial V}\right)_T = -\frac{nRT}{V^2} = 0 \tag{4}$$

$$\left(\frac{\partial^2 p}{\partial V^2}\right)_T = -\frac{2nRT}{V^3} = 0 \tag{5}$$

It is easy to see that the only way these two equations can be satisfied is if $T = 0$, or $V = \infty$ Neither of these solutions is physically reasonable so we conclude that the ideal gas does not have a critical point.

Good exercises would be for you to see if the approximate equation of state,

$$p(V - nb) = nRT$$

has a critical point, or to verify for yourself that the van der Waals equation of state does have a critical point and to find the critical constants, V_C, T_C, and p_C.

CRITICAL CONSTANTS OF THE VAN DER WAALS GAS

We saw in our discussion of critical phenomena that the mathematical definition of the critical point is,

$$\left(\frac{\partial p}{\partial V}\right)_T = 0 \tag{1}$$

and

$$\left(\frac{\partial^2 p}{\partial V^2}\right)_T = 0 \tag{2}$$

In other words, the critical isotherm on a p-V diagram has a point of inflection. Equations (1) and (2) constitute a set of two equation in two unknowns, V, and T. One can test to see if an approximate equation of state gives a critical point by calculating these two derivatives for the equation of state and trying to solve the pair of equations. If a solution exists (and T and V are neither zero or infinity) then we say that the equation of state has a critical point.

Let's use this test to see if a van der Waals gas has a critical point. First we have to solve the van der Waals equation of state for pressure, p,

$$p = \frac{nRT}{V - nb} - a\frac{n^2}{V^2} \tag{3}$$

Now we can take the derivatives in Equations 1 and 2 and set them (independently) equal to zero.

$$\left(\frac{\partial p}{\partial V}\right)_T = -\frac{nRT}{(V - nb)^2} 2a\frac{n^3}{V^3} = 0 \tag{4}$$

$$\left(\frac{\partial^2 p}{\partial V^2}\right)_T = -\frac{2nRT}{(V - nb)^3} - 6a\frac{n^2}{V^4} = 0 \tag{5}$$

In order to stress that from here on the problem is pure algebra, let's rewrite the simultaneous equations that must be solved for the two unknowns V and T (which solutions we will call V_C and T_C),

$$-\frac{nRT}{(V - nb)^2} + 2a\frac{n^2}{V^3} = 0 \tag{6}$$

$$-\frac{2nRT}{(V - nb)^3} + 6a\frac{n^2}{V^4} = 0 \tag{7}$$

There are several ways to solve simultaneous equations. One way is to multiply Equation (6) by,

$$\frac{2}{V - nb}$$

to get

$$-\frac{2nRT}{(V-nb)^3}+4a\frac{n^2}{V^3(V-nb)}=0 \tag{8}$$

Now add equations (7) and (8). Note that in this addition the terms containing T will cancel out leaving,

$$-\frac{6an^2}{V^4}+4a\frac{n^2}{V^3(V-nb)}=0 \tag{9}$$

Divide Equation (9) by $2an^2$ and multiply it by V^3 (and bring the negative term to the other side of the equal sign) to get,

$$\frac{2}{(V-nb)}=\frac{3}{V}, \tag{10}$$

which is easily solved to get

$$V=V_C=3nb. \tag{11}$$

To find the critical temperature, substitute the critical volume from Equation (11) into one of the derivatives (which equals zero) say Equation (6). This gives,

$$-\frac{nRT}{(3nb-nb)^2}+2a\frac{n^2}{(3nb)^3}=0, \tag{12}$$

which "cleans up" to give,

$$\frac{RT}{4}=\frac{2a}{27b}, \tag{13}$$

or

$$T=Tc=\frac{.8a}{27bR} \tag{14}$$

The critical pressure is obtained by substituting V_C and T_C into the van der Waals equations of state as solved for p in Equation (3).

$$Pc=\frac{nRT_C}{V_C-nb}-a\frac{n^2}{V_C^2}$$

$$Pc = \frac{n\left(\frac{8a}{27bR}\right)}{3nb - nb} - a\frac{n^2}{(3nb)^2} \tag{15 a,b}$$

This simplifies to,

$$Pc = \frac{a}{27b^2} \tag{16}$$

Our conclusion is that the van der Waals equation of state does give a critical point since the set of simultaneous equations (Equations (1) and (2)) has a unique solution.

The van der Waals equation of state is still an approximate equation of state and does not represent any real gas exactly. However, it has some of the features of a real gas and is therefore useful as the next best approximation to a real gas. We will be deriving thermodynamic relationships (equations) using the ideal gas approximation. We can rederive some of these equations using the van der Walls equation of state in order to see how these relationships are affected by gas nonideality. .

Chapter 4

Solids and Liquids

There are approximate equations of state for gases which can give virtually any degree of accuracy desired. However, there are no analogous equations of state for solids and liquids. Fortunately the volumes of solids and liquids do not change very much with pressure as long as the pressure changes are not too large. This situation allows us to define parameters and form an approximate equation of state which is valid over a moderate range of temperatures and pressures.

We will restrict our attention to isotropic liquids and solids, which means that we are excluding liquid crystals and solid single crystals. Single crystals and liquid crystals are anisotropic. Their response to pressure and their expansion with temperature is different along different axes in the crystal. (Many solids, particularly metals and alloys are conglomerates of microscopic crystals with random orientations so that the bulk material behaves like an isotropic solid even though the individual microscopic crystals are anisotropic. We can apply our methods for isotropic substances to these materials even though, strictly speaking, they are crystalline.)

The volume of a sample of an isotropic material is known experimentally to be a function of temperature and pressure. Therefore, we can write,

$$V = V(T, p) \tag{1}$$

(The volume is also a function of the number of moles in the sample, but we will be looking at relative changes, or fractional changes, so that the quantity of material will cancel out.)

We write a differential change in the volume due to differential changes in the temperature and/or the pressure as follows:

$$dV = \left(\frac{\partial V}{\partial p}\right)_T dp + \left(\frac{\partial V}{\partial p}\right)_p dT \tag{2}$$

The relative change, or fractional change, is then,

$$\frac{dV}{V} = \frac{1}{V}\left(\frac{\partial V}{\partial p}\right)_T dp + \frac{1}{V}\left(\frac{\partial V}{\partial p}\right)_p dT \tag{3}$$

The coefficients of *dp* and *dT* in Equation (3) are so important that we give then names and special symbols.

$$K_T = \frac{1}{V}\left(\frac{\partial V}{\partial p}\right)_T \tag{4}$$

is called the isothermal compressibility. The subscript, *T*, on the lower case Greek letter kappa is to distinguish this compressibility from another related one which will be defined later. When the pressure is increased the volume decreases so that the derivative in Equation (4) is negative. The negative sign in the definition of k_T ensures that kappa is positive When there is no concern about confusion we will omit the subscript on the kappa.

$$\alpha = \frac{1}{V}\left(\frac{\partial V}{\partial p}\right)_p \tag{5}$$

is called the coefficient of thermal expansion (or sometimes just the expansion coefficient).

Values of α and k_T must be obtained from experimental data and they can be found in data tables. α and k_T are themselves functions of temperature and pressure although they vary so slowly with temperature and pressure that they may usually be regarded as constants except over very large temperature or pressure intervals. We will regard them as constant.

Although α and k_T are most useful for liquids and solids, they can be calculated for gases. The volume of a gas is a strong function of temperature and pressure so α and k_T are not even approximately constant for gases. It is a useful exercise in the application of partial derivatives to calculate these quantities for an ideal gas. For example, using the ideal gas equation of state we get,

$$V = \frac{nRT}{P},$$

so that

$$\alpha = \frac{1}{V}\left(\frac{\partial V}{\partial T}\right)_p$$

$$= \frac{1}{V}\left(\frac{nR}{p}\right) = \frac{nR}{pV}$$

$$= \frac{1}{T} \quad (6,a,b,c)$$

This quantity is clearly not constant. (Bear in mind that this is the coefficient of thermal expansion for an ideal gas, not the general expression for α. We will leave it to the reader to show that the isothermal compressibility for an ideal gas is $1/p$).

Equation (3) can be rewritten, using α and k_T as

$$\frac{dV}{V} = -K_r dp + \alpha dT, \quad (7)$$

which can be integrated to give an approximate equation of state for isotropic liquids and solids,

$$V = V_o e^{-k_r(p-p_o)+\alpha(T-T_o)}, \quad (8)$$

where V_0 is the volume at p_0 and T_0. It is a useful exercise for the reader to show that this approximate equation of state is consistent with our definitions of α and k_T.

There is one other quantity of interest which can be obtained from α and k_T, namely,

$$\left(\frac{\partial p}{\partial T}\right)_V .$$

This is the derivative that tells us how fast the pressure rises when we try to keep the volume constant while increasing the temperature. Using a variation of Euler's chain rule we can write,

$$\left(\frac{\partial p}{\partial T}\right)_V = -\frac{\left(\frac{\partial V}{\partial T}\right)_p}{\left(\frac{\partial V}{\partial p}\right)_T}$$

$$= -\frac{\frac{1}{V}\left(\frac{\partial V}{\partial T}\right)_p}{\frac{1}{V}\left(\frac{\partial V}{\partial p}\right)_T}$$

$$= \frac{\alpha}{k_T} \cdot \qquad (9,a,b,c)$$

Let's apply this to see how much pressure would be generated in a mercury thermometer if we tried to heat the thermometer higher than the temperature where the mercury has reached the top of the thermometer.. For Hg, $\alpha = 1.82 \times 10^{-4}\ K^{-1}$ and $k_T = 3.87 \times 10^{-5}\ atm^{-1}$. We write.

$$dp = \frac{\alpha}{k_T}$$

$$= \frac{1.82 \times 10^{-4} K^{-1}}{3.87 \times 10^{-5} atm^{-1}}$$

$$= 4.70\, atm / K dT. \qquad (10,a,b,c)$$

So we see that each 1 °C increases the pressure by 4.7 atm, about 69 lb/sq in. This is a lot of pressure for a glass tube to withstand. It wouldn't take very many degrees of temperature increase to break the glass thermometer.

THERMOMETERS AND THE IDEAL GAS TEMPERATURE SCALE

Many of the thermometers we see and use are made of a thin glass tube containing a liquid. The temperature is measured by observing how far up the tube the liquid rises. However, we have already seen that α is not a constant so that liquid expansion is not uniform and the rise in the liquid is not linear with temperature. Worse, different liquids have different nonlinear expansions.

We could pick a standard substance and all agree to measure temperature by the expansion of this substance, but it is unsatisfactory to have our measuring devices tied to particular substances. It would be best if we had a temperature measuring device which was independent of any particular material.

The ideal gas thermometer is such a device and the temperature scale it defines is called the ideal gas temperature scale. The ideal gas temperature scale is based on the fact that all gases become ideal in the limit of zero pressure. Therefore, we can define the ideal gas temperature as,

$$T = \frac{1}{nR}\lim_{p\to 0}\left(p^{V}\right). \tag{11}$$

This temperature scale is independent of the gas used. It has a natural zero since $p > 0$ and $V > 0$, so that pV is never negative. The value of R determines the size of the degree. If R is the gas constant, 0.082057459 Latm/Kmol, then the degree is the Kelvin degree. No one claims that the ideal gas thermometer is easy to use, but it does provide us with an unambiguous theoretical standard to establish a temperature scale.

ENERGY, WORK, AND HEAT

ENERGY AND WORK

Thermodynamics deals with energy in its various forms and the conversion of one form of energy into another. Energy appears in several different forms, kinetic energy, potential energy, heat, chemical energy, and so on.

Kinetic energy is energy of motion. It is written,

$$KE = \frac{1}{2}mv^2 \tag{1}$$

where m, is the mass of a moving object and v is its velocity.

Potential energy has many different forms, depending on the physical system at hand. For example,

mgh- local gravity,

$\frac{1}{2}kx^2$ - Hook's law, compression of a spring,

$-\frac{e^2}{r}$ - Coulomb's law,

$-\frac{mMG}{r}$ - large scale gravity,

and so on.

Kinetic and potential energy are interchangeable and both can be converted into work.

Thermodynamics does not provide us with the expressions for kinetic energy, potential energy, or work. These must come from physics.

Mechanical work (from physics) is a force times the distance through which it acts. That is,

$$dW = f\,dx, \tag{2}$$

for one dimensional motion.

Work for a finite motion is obtained by integrating Equation (2),

$$w = \int_{x_1}^{x_2} f(x)dx, \tag{3}$$

where the $f(x)$ takes into account the possibility that the force may be changing as one moves along the path from x_1 to x_2.

Work can increase the kinetic or potential energy of a system.

HEAT

One of the great breakthroughs in the history of science was the recognition that heat is a form of energy. Since it was known that heat "flowed" from a hot body to a cold body heat was thought to be a fluid of some sort - called phlogiston. When experiment showed that the products of combustion weighed more that the object combusted, and yet the combustion process gave off heat, it was necessary to make the unlikely assertion that phlogiston had a negative mass.

Benjamine Thompson, also known as Count Rumford of the Holy Roman Empire discovered the true nature of heat as a form of energy while operating a factory for boring cannon. In the process of boring the hole in the barrel of a cannon the metal got hot. Rumford was able to show that the only explanation for this phenomenon was that the work being put into turning the drill bit was being converted into heat. He even made an attempt to determine the "mechanical equivalent of heat." Joule later improved on his measurements and obtained a value close to the modern one of 4.184 J = 1 cal (in modern units).

The conclusion is that heat is a form of energy. A revolutionary conclusion for its time, but no big surprise now.

Work, kinetic energy, and potential energy can be converted into heat with no restrictions.

Heat can be converted into work, kinetic energy, and potential energy, but only with restrictions (which we will discuss in due time).

DEFINITIONS AND CONVENTIONS

We define the system as the object or sample or "thing" we are interested in. The surroundings is everything else. For a given thermodynamics discussion we can say,

system + surroundings = the universe.

We define w as the work done on the system, and q as the heat absorbed by the system. This means that w and q are algebraic quantities. They can be either positive or negative and their sign tells us which way energy is flowing. For example, if w is positive it means that work was done on the system so that the energy of the system increased, and so on. Likewise, if q is negative the system lost heat to the surroundings.

(In older books w was defined as the work done on the surroundings. There is a reason for this. It is sometimes easier to calculate the work done on the surroundings than to calculate work done on the system. Nevertheless, modern books use the convention given above that w is work done on the system. If you are reading an older book and there seems to be a sign error, it may be because they are using the older convention for w.)

We will define w' as work done on the surroundings. Clearly,

$$w' = -w.$$

We now define a quantity called the internal energy, U. The name of the variable, U, is self explanatory. U is the total energy contained in the system.

(Thermodynamics does not care whether or not there are atoms and molecules. Everything that we do in thermodynamics can be done without ever knowing that there are atoms and molecules. However, just to calibrate our intuition, it may be useful to say that the internal energy is the sum of all the kinetic and potential energies of all the particles in the system. We will define three other thermodynamic variables or functions which have units of energy, but none of these will have a simple description such as we have for U.

One other comment. A measurement of energy depends on where you measure the energy from. For example, the potential energy of a person standing on the surface of the earth might be considered to be zero relative to local gravity - mgh - but would be large and negative relative the large scale gravitational system such as the earth-moon system.)

THE FIRST LAW OF THERMODYNAMICS

There are several word statements of the first law of thermodynamics:

- Energy is conserved.
- (Which is another way of saying that energy cannot be created or destroyed. You can change its form, but you cannot create it or destroy it.)
- It is impossible to make a perpetual motion machine of the first kind.

(A perpetual motion machine of the first kind is a system that gives energy to the surroundings, but produces no change in the system itself and no other change to the surroundings. This statement implies that there is a perpetual motion machine of the second kind. We will find out about a perpetual motion machine of the second kind when we meet the second law of thermodynamics.)

The mathematical statement of the first law is phrased in terms of a process. Given any change or process,

initial state • final state

$\Delta U = U_{\text{final}} - U_{\text{initial}},$

or

state 1 – state 2

$\Delta U = U_2 - U_1$.

(Initial and final states must both be at equilibrium.)

Then the first law of thermodynamics says that

$\Delta U = q + w.$

The first law of thermodynamics is a law of observation. No one has ever observed a situation where energy is not conserved so we elevate this observation to the status of a law. The real justification of this comes when the things we derive using the first law turn out to be true - that is, verified by experiment.

(Actually there are situations were energy is not conserved. We now know that in processes where the nuclear structure of matter is altered mass can be converted into energy and vice versa. This is a consequence of special relativity were it is found that matter has a "rest energy," mc^2, where m is the mass to be converted to energy and c is the speed of light. As a consequence of nuclear energy we should say that,

Energy + the energy equivalent of mass is conserved.

Then the first law would be written,

$\Delta U = q + w + \Delta mc^2.$

For chemical processes the change in energy due to changes in mass is negligible - though not zero - so we can ignore it.)

The first law can be written in differential form,

$dU = dq + dw$

Which is called the differential form of the first law.

(Actually, this is the differential form of the first law for a closed system, that is, for a system in which no material moves in or out of the system. Later we will write the differential form of the first law for an open system, where material can move in or out of the system.)

Note: Some writers like to use a special symbol for the d in dq and dw to indicate that these differentials are not in the same mathematical class as, for example, dU. We will not use this notation. As soon as we have learned what the difficulty is with the present d you will be expected just to remember that the d in dq and dw is different than the d in dU.

PV WORK

We have seen that the expression for work must be obtained from physics. The expression for mechanical work, force times distance, is given by,

$$dW = f dx,$$

or, for a finite change,

$$W = \int_{x_1}^{x_2} f(x)\,dx.$$

We would now like to apply these expressions for mechanical work to the case where work is accomplished by the expansion or contraction of a system under an external pressure.

Let us consider a cylinder of cross-sectional area A fitted with a piston. The apparatus is arranged so that the piston encloses a sample at pressure $p_{int,}$ and the piston is attached to a mechanism which will maintain an external pressure, p_{ext}, in the apparatus. We will assume that $p_{int} \geq p_{ext}$.

It turns out that it is easier to calculate the work done on the surroundings, w'. (Recall that $w' = -w$.) In this case,

$$dw' = f dx. \qquad (1)$$

The piston is released to move a distance dx. Since pressure is force per unit area, the force against which the piston moves is $p_{ext} A$. So the work, dw' is

$$dw' = f dx = p_{ext} A dx. \qquad (2)$$

But Adx is a differential volume swept out by the piston in the expansion. Call the differential volume $A dx = dV$. Then

$$dw' = f dx = p_{ext} A dx = p_{ext} dV. \qquad (3)$$

Going back to work done on the system, dw, we find,

$$dw = -dw' = -p_{ext}\,dV. \tag{4}$$

REVERSIBLE AND IRREVERSIBLE PROCESSES

A reversible process is one that can be halted at any stage and reversed. In a reversible process the system is at equilibrium at every stage of the process. An irreversible process is one where these conditions are not fulfilled.

If $p_{int} > p_{ext}$ in an expansion process then the process is irreversible because the system does not remain at equilibrium at every stage of the process. (There will be turbulence and temperature gradients, for example.) For irreversible processes, pV work must be calculated using

$$dw = -p_{ext}dV. \tag{5}$$

On the other hand, if $p_{int} = p_{ext}$ then the process can be carried out reversibly. Also, there is then no need to distinguish between external pressure and internal pressure so that

$$p_{int} = p_{ext} = p$$

and there is only one pressure defined for the system. In this case, which will account for the majority of problems that we deal with,

$$dw = -pdV, \tag{6}$$

and

$$w = -\int_{V_1}^{V_2} p\,dV. \tag{7}$$

Example Calculations

First example: A reversible expansion with $dp = 0$. That is, a process at constant pressure.

We write our expression for reversible work done on the system,

$$w = -\int_{V_1}^{V_2} p\,dV. \tag{7}$$

If pressure is constant then the p can be brought outside the integral to give,

$$w = -\int_{V_1}^{V_2} p\,dV$$

$$= -p\int_{V_1}^{V_2} dV$$

$$= -p(V_2 - V_1)$$

$$= -p\Delta V. \qquad (8, a, b, c, d)$$

(The answer will come out in Latm and should be converted to J using 1 Latm = 101.325 J.

Second example: An isothermal reversible expansion. That is, $dT = 0$. We use the same starting place

$$w = \int_{V_1}^{V_2} p\,dV, \qquad (7)$$

but this time pressure is not constant and will change as V changes,

$$w = -\int_{V_1}^{V_2} p(V)\,dV. \qquad (9)$$

In order to do the integration we must know how pressure varies with volume. We can obtain this information from the equation of state. If our substance is a gas we can get an approximate value of the expansion work using the ideal gas equation of state, where

$$p = \frac{nRT}{V}. \qquad (10)$$

Substituting the ideal gas expression for pressure into Equation (7) we get

$$w = -\int_{V_1}^{V_2} \frac{nRT}{V}\,dV. \qquad (11)$$

This time T is constant so that we can bring the nRT outside the integral to get.

$$w = -nRT \int_{V_1}^{V_2} \frac{1}{V} dV, \tag{12}$$

Which integrates to give

$$w = -nRT(In\, V_2 - V_1)$$

$$= -nRT\, In \frac{V_2}{V_1}. \tag{13a, b}$$

The next best approximation would be to approximate the volume dependence of the pressure using the van der Waals equation of state.

$$P = \frac{nRT}{V - nb} - \frac{an^2}{V^2}.$$

We will leave it as an exercise for the reader to calculate the expansion work for a van der Waals gas.

A better approximation yet could be obtained using the virial expansion to give the volume dependence of pressure,

$$P = \frac{nRT}{V}\left(1 + \frac{B(T)n}{V} + \frac{C(T)n^2}{V^2} + ...\right).$$

The General Case

Suppose we go from p_1V_1 to p_2V_2 by some general path. The reversible work is still represented by Equation (7),

$$w = -\int_{V_1}^{V_2} pdV. \tag{7}$$

The path from p_1V_1 to p_2V_2 can be represented by a curve on a p-V diagram.

The integral in Equation (7) can be represented by the area under the curve which goes from p_1V_1 to p_2V_2, so that the work becomes,

$w =$ " area.

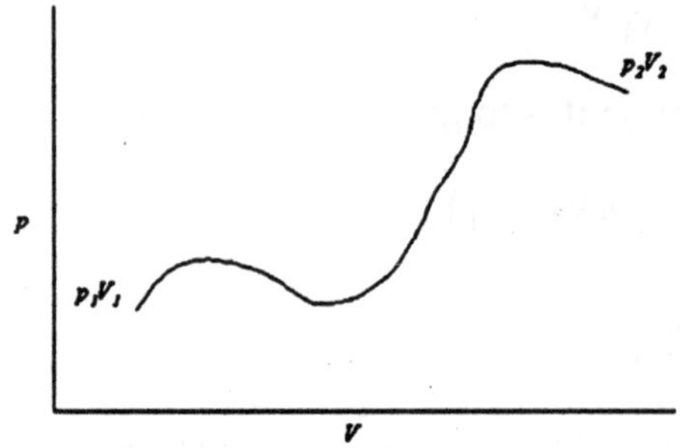

Notice that there are many possible curves which would connect the points p_1V_1 and p_2V_2 and each curve would have a different area and give a different value for w. We conclude that w depends on the path, unlike ΔU which only depends on the initial and final states. We call variables like U, p, V, T, and so on, state variables because ΔU, Δp, ΔV, ΔT, and so on, do not depend on the path, but only on the initial and final states of the system. A quantity like w which does depend on path is not a state variable. We will never write w with a Δ in front of it.

We will soon see that q is also path dependent.

HEAT AND HEAT CAPACITY

If we add heat to a sample of material, often the temperature will increase. (If we are at the temperature of a phase change, for example ice in water, the temperature will not change it will just melt some of the ice.) Away from a phase change adding heat will always give an increase in temperature. The amount of the temperature increase depends on how much heat was added, the size of the sample, the original temperature of the sample, and on how the heat was added. The two obvious choices on how to add the heat are to add it holding volume constant or to add it holding pressure constant.

Let's assume for the moment that we are going to add heat to our sample holding volume constant, that is, $dV = 0$. Let q_V be the heat added[1] (the subscript, V, indicates that the heat is being added at constant V). Also, let ΔT be the temperature change. The ratio, $\frac{qv}{\Delta T}$, depends on the material, the amount of material, and the temperature. In the limit where q_V goes to zero (so that ΔT also goes to zero) this ratio becomes a derivative,

$$\lim_{q_r \to 0}\left(\frac{qV}{\Delta T}\right)_V = \left(\frac{\partial q}{\partial T}\right)_V = C_V \tag{1}$$

We have given this derivative the symbol, C_V, and we call it the "heat capacity at constant volume. Usually one quotes the "molar heat capacity,"

$$\overline{C}_V = C_{Vm} = \frac{C_V}{n} \tag{2}$$

We can rearrange Equation 1 as follows,

$$dq_V = C_V dT. \tag{3}$$

Then we can integrate this equation to find the heat involved in a finite change at constant volume,

$$q_V = \int_{T_1}^{T_2} C_V dT \tag{4}$$

If C_V is approximately constant over the temperature range then C_V comes out of the integral and the heat at constant volume becomes,

$$q_V = C_V\left(T_2 - T_1\right). \tag{5}$$

Let us now go through the same sequence of steps except holding pressure constant instead of volume. Our initial definition of the heat capacity at constant pressure, C_p becomes,

$$\lim_{q_p \to 0}\left(\frac{q_p}{\Delta T}\right)_p = \left(\frac{\partial q}{\partial T}\right)_p = C_p \tag{6}$$

The analogous molar heat capacity is,

$$\overline{C}_p = C_{pm} = \frac{C_p}{n} \tag{7}$$

Equation (6) rearranges to,

$$dq_p = C_p dT \tag{8}$$

which integrates to give,

$$q_p = \int_{T_1}^{T_2} C_p dT \tag{9}$$

When C_p is approximately constant the integral in Equation (9) becomes

$$q_p = C_p (T_2 - T_1) \tag{10}$$

Very frequently the temperature range is large enough that C_p cannot be regarded as constant. In these cases the heat capacity is fit to a polynomial (or similar function) in T. For example, some tables give the heat capacity as,

$$\overline{C}_p = \alpha + \beta T + \gamma T^2 \tag{11}$$

where α, β and γ are constants given in the table. With this temperature-dependent heat capacity the heat at constant pressure would integrate as follows,

$$q_p = n \int_{T_1}^{T_2} (\alpha + \beta T + \gamma T^2) dT$$

$$= n\alpha\left(T_2 - T_2\right) + n\frac{\beta}{2}\left(T_2^2 - T_2^2\right) + n\frac{\gamma}{3}\left(T_2^3 - T_1^3\right) \tag{12a, b}$$

Occasionally one finds a different form for the temperature dependent heat capacity in the literature,

$$\overline{C}_p = a + bT + cT^{-2} \tag{13}$$

When you do calculations with temperature dependent heat capacities you must check to see which form is being used for C_p.

1. We are using the convention that q will always designate heat absorbed by the system. q can be positive or negative and the sign indicates which way heat is flowing. If q is positive then heat was indeed absorbed by the system. On the other hand, if q is negative it means that the system gave up heat to the surroundings.

ENERGY, THE FIRST LAW, AND ENTHALPY

We have agreed that work, potential energy, kinetic energy, and heat are all forms of energy. Historically, it was not obvious that heat belonged in this list. But beginning with the experiments of Count Rumford of the Holy Roman Empire, and later the experiments of Joule, it became clear that heat, too, was just another form (or manifestation) of energy.

Recall that we defined the internal energy, U, as the total energy of the system. (Although the existence of atoms and molecules is not relevant to thermodynamics, we said that the internal energy is the sum of all the kinetic and potential energies of all the particles in the system. This statement is outside the realm of thermodynamics, but it is useful for us to gain an intuitive "feel" for what the internal energy is.)

Recall also that energies are always measured relative to some origin of energy. The origin is irrelevent to thermodynamics because we will always calculate changes in U and not absolute values of U. That is, we calculate

$$\Delta U = U_{final} - U_{initial} \tag{1}$$

In words, this equation reads, "the change in the internal energy is equal to the final internal energy minus the initial internal energy." This equation also reminds us that U is a "state function." That is, the change in U does not depend on how the change was done (in other words, on the path), but depends only on the initial and final states.

Recall that the first law of thermodynamics in equation form for a finite change, is given by,

$$\Delta U = q + w \tag{2}$$

Equation (2) tells something else of importance. We know that U is a state function and that ΔU is independent of path. However, w is not a state function so that w depends on path. Yet the sum of w and q is path independent. The only way this can happen is if q is also path dependent. We now see that we are dealing with two path-dependent quantities, q and w.

For a differential change we write the first law in differential form,

$$dU = dq + dw \tag{3}$$

The w in Equation (2) or the dw in Equation (3)3 includes all types of work, work done in expansion and contraction, electrical work, work done in creating new surface area, and so on. Much of the work that we deal with in thermodynamics will be work done in expansion and contraction of the system, or pV work. Recall that the expression for pV work is,

$$dw = -pdV \tag{4}$$

If we want to include both pV work and other types of work we can write the first law as,

$$dU = dq - pdV + dw_{other} \tag{5}$$

Let's now confine ourselves to systems where there is only pV work. In this case the first law can be written,

$$dU = dq - pdV \tag{6}$$

Suppose we now regard U as a function of T and V. That is, $U = U(T,V)$. Then, for dU we can write,

$$dU = \left(\frac{\partial U}{\partial T}\right)_V dT + \left(\frac{\partial U}{\partial V}\right)_T dV \tag{7}$$

For a process at constant V ($dV = 0$) Equations (6) and (7) become,

$$dU_V = dq_V \tag{8}$$

and

$$dU_V = \left(\frac{\partial U}{\partial T}\right)_V dT \tag{9}$$

We know, from our discussion on heat and heat capacity, that the differential heat at constant volume can also be written as,

$$dq_V = C_V dT \tag{10}$$

so,

$$dU_V = C_V dT \tag{11}$$

Comparing Equations (9) and (11), and recognizing that the change dU_V is the same in both cases, we see that,

$$C_V = \left(\frac{\partial U}{\partial T}\right)_V \tag{12}$$

We shall regard Equation (12) as the formal thermodynamic definition of the heat capacity at constant volume. This new definition is more satisfactory than our previous temporary definition,

$$C_V = \left(\frac{\partial q}{\partial T}\right)_V \tag{13}$$

Equation (12) is a better definition of the heat capacity because it is usually more satisfactory to define thermodynamics quantities in terms of state functions, like U, T, V, p, and so on, rather than on things like q and w which depend on path.

One other comment, we can integrate Equation (8), at constant volume, to get,

$$\Delta U_V = q_V \tag{14}$$

In words, for any process at constant volume the heat, q, is the same as the change in the internal energy, ΔU.

ENTHALPY

It turns out that V is not the most convenient variable to work with or to hold constant. It is much easier to control the pressure, p, on a system than it is to control the volume of the system, especially if the system is a solid or a liquid. What we need is a new function, with units of energy, which contains all the information that is contained in U but which can be controlled by controlling the pressure. Such a function can be defined (created) by a Legendre transformation. There are particular criteria which must be met in making a Legendre transformation, but in our case here these criteria are met. (A full discussion of the mathematical properties of Legendre transformations is beyond the scope of this discussion. There are more details given in the Appendices to Alberty and Silby.) In our case we will define a new quantity, H, called the enthalpy, which has units of energy, as follows,

$$H = U + pV \tag{15}$$

We can show that H is a natural function of p (in the same sense that U is a natural function of V) as follows,

$$\begin{aligned} dH &= dU + pdV + Vdp \\ &= dq - pdV + pdV + Vpd \\ &= dq + Vdp \end{aligned} \tag{16a, b, c}$$

One of the great utilities of the enthalpy is that it allows us to use a state function, H, to describe the heat involved in processes at constant pressure rather than the heat, q, which is not a state function. To see this, let's go through the same process with dH that we did with dU above. Let's regard H as a function of T and p (for now). Then we can write,

$$dH = \left(\frac{\partial H}{\partial T}\right)_p dT + \left(\frac{\partial H}{\partial p}\right)_T dp \tag{17}$$

Consider a process at constant pressure ($dp = 0$). From Equation (16c) we conclude that.

$$dH_p = dq_p \tag{18}$$

and from Equation (17) we get,

$$dH_p = \left(\frac{\partial H}{\partial T}\right)_p dT \tag{19}$$

We know, from our discussion on heat and heat capacity, that the differential heat at constant pressure can also be written as,

$$dq_p = C_p dT \tag{20}$$

so,

$$dH_p = C_p dT \tag{21}$$

Comparing Equations (19) and (21), and recognizing that the change dH_p is the same in both cases, we see that,

$$C_p = \left(\frac{\partial H}{\partial T}\right)_p \tag{22}$$

We shall regard Equation (22) as the formal thermodynamic definition of the heat capacity at constant pressure. Again, this definition is much more satisfactory than our previous temporary definition,

$$C_p = \left(\frac{\partial q}{\partial T}\right)_p \tag{23}$$

since it defines the heat capacity in terms of the state function, H, rather than in terms of q which is not a state function.

Just as we integrated equation (8), we can integrate Equation (21), at constant pressure, to get,

$$\Delta H_p = q_p \tag{24}$$

In words, for any process at constant pressure the heat, q, is the same as the change in enthalpy, ΔH. This equation contains no approximations. It is valid for all process at constant pressure. Equation (24) is vastly more useful than its counterpart at constant volume because we carry out our chemistry at constant pressure much more often than we do at constant volume.

People sometimes ask, "What is the meaning of H?" Unfortunately, there is no simple, intuitive physical description of enthalpy like there is for the internal energy. (Recall that the internal energy is the sum of all kinetic and potential energies of all the particles in the system). The nearest thing we can come to as a description of H is the one above where ΔH is the heat (gain or loss) in a constant pressure process. For this reason the enthalpy is ocassionally referred to as the "heat content."

Nuclear energy was unknown to the original formulators of thermodynamics. We now know that matter can be converted into energy and vice versa. The "energy equivalent of matter" is given by the famous Einstein formula, $E = mc^2$, where m is the mass of the matter and c is the velocity of light. Since the velocity of light is very large, about 3 x 10^8 m/s, a small amount of mass is equivalent to a very large amount of energy. Strictly speaking, the statement, "energy is conserved," should be replaced by the statement, "energy plus the energy equivalent of mass is conserved." That is, energy + mc^2 is conserved. The conversion of mass to energy or energy to mass in chemical reactions is so small that it is virtually never observed in chemical problems. So, for chemical thermodynamics, the simpler statement that energy is conserved is sufficient

THE JOULE EXPANSION

Much of the early progress in thermodynamics was made in the study of the properties of gases. One of the early questions was whether or not gases cool on expansion. (Our intuition might tell us that they would, but is our intuition correct?)

Joule designed an experiment to find out whether or not gases cool on expansion and if so how much.

The Joule apparatus consisted of two glass bulbs connected by a stopcock. One bulb was filled with gas at some p and T. The other bulb was evacuated. The entire apparatus was insulated so that $q = 0$. That is, the experiment would be adiabatic.

The stopcock was opened to allow the gas to expand into the adjoining bulb. Since the gas was expanding against zero pressure no work was done, $w = 0$. With both $q = 0$ and $w = 0$ it is clear that,

$$\Delta U = q + w = 0.$$

The process is at constant internal energy.

Clearly, $\Delta V \neq 0$ because the gas expanded to fill both bulbs.The question was, did T change? ΔT was measured to be zero, no temperature change.

(It turns out that the Joule experiment was sufficiently crude that it could not detect the difference between an ideal gas and a real gas so that the conclusions we will draw from this experiments only apply to an ideal gas.)

In effect, Joule was trying to measure the derivative,

$$\left(\frac{\partial p}{\partial T}\right)_U,$$

and the result was that,

$$\left(\frac{\partial p}{\partial T}\right)_U = 0 \tag{1}$$

This particular derivative is not all that instructive, with U being held constant. We can use our version of Euler's chain relation to obtain information that is more instructive.

$$\left(\frac{\partial T}{\partial V}\right)_U = \frac{\left(\frac{\partial U}{\partial V}\right)_T}{\left(\frac{\partial U}{\partial V}\right)_V} = \frac{\left(\frac{\partial U}{\partial V}\right)_T}{C_V}$$

$$= 0 \tag{2a, b, c}$$

We know that C_V for gases is neither zero nor infinity, so we must conclude that,

$$\left(\frac{\partial U}{\partial V}\right)_V = 0 \tag{3}$$

This is an important and useful result. It says that the internal energy of an ideal gas is not a function of T and V, but of T only. That is, in equation form,

$$\text{for an ideal gas } U = U(T). \tag{4}$$

For real gases, and most approximations to real gases, like the van der Waals equation of state,

$$\left(\frac{\partial U}{\partial V}\right)_V \neq 0$$

However, this quantity is quite small, even for real gases. We will have occasion to calculate it for the van der Waals equation of state later on.

This result extends to the enthalpy of an ideal gas.

$$H = U + pV = U(T) + nRT = H(T). \quad (5)$$

Thus, for an ideal gas both U and H are functions of T only.

Then all of the following derivatives are zero:

$$\left(\frac{\partial U}{\partial V}\right)_T = 0,$$

$$\left(\frac{\partial U}{\partial p}\right)_T = 0,$$

$$\left(\frac{\partial H}{\partial V}\right)_T = 0,$$

$$\left(\frac{\partial U}{\partial p}\right)_T = 0. \quad \text{(6a, b, c, d)}$$

We will now use some of these results to discuss that adiabatic expansion of an ideal gas.

ADIABATIC EXPANSION OF AN IDEAL GAS

The definition of an adiabatic expansion, for now, is $dq = 0$. That is, no heat goes in or out of the system. However, $dw \neq 0$. As the gas expands it does work on the surroundings. Since the gas is cut off from any heat bath it can not draw heat from any source to convert into work. The work must come from the internal energy of the gas so that the internal energy decreases. Since the internal energy of an ideal gas in only dependent on T that means that the temperature of the gas must decrease.

From the first law with only pV work we have

$$dU = dq - pdV$$

$$= -pdV \tag{7a, b}$$

because $dq = 0$ for an adiabatic process.

Regarding U as a function of T and V. That is, $U = U(T,V)$, we get

$$dU = \left(\frac{\partial U}{\partial T}\right)_V dT + \left(\frac{\partial U}{\partial V}\right)_r dV$$

$$C_V dT, \tag{8a, b}$$

because of the definition of C_V and because our gas is an ideal gas so that the second derivative vanishes (Equation (6a)).

The dU's in Equations (7) and (8) must be equal so that

$$-pdV = C_V dT$$

$$-\frac{nRT}{V}dV = C_V dT. \tag{9a, b}$$

Rearranging Equation (9b) we get

$$\frac{dT}{T} = \frac{nR}{C_V}\frac{dV}{V}. \tag{10}$$

By the same token, using enthalpy, we find

$$dH = dq + Vdp$$

$$= Vdp, \tag{11a, b}$$

and

$$dH = \left(\frac{\partial H}{\partial T}\right)_p dT + \left(\frac{\partial H}{\partial p}\right)_r dp$$

$$= C_p dT, \tag{12a, b}$$

From which we deduce that

$$\frac{dT}{T} = \frac{nR}{Cp}\frac{dp}{p}. \tag{13}$$

Comparing Equations (10) and (13) we see that

$$-\frac{nR}{C_V}\frac{dV}{V}=\frac{nR}{C_p}\frac{dp}{p}, \qquad = -\frac{C_p}{C_V}\frac{dV}{V}=\frac{dp}{p}$$

$$-\gamma\frac{dV}{V}=\frac{dp}{p}, \qquad \text{(14a, b, c)}$$

Where we have written $C_p/C_V = \gamma$.

If we regard C_p and C_V as constant then Equation (14c) can be integrated to give,

$$-\gamma In\frac{V_2}{V_1}=In\frac{p_2}{p_1},$$

$$\gamma In\frac{V_2}{V_1}=In\frac{p_2}{p_1},$$

$$In\left(\frac{V_1^\gamma}{V_2^\gamma}\right)=In\frac{p_2}{p_1};$$

$$\frac{V_1^\gamma}{V_2^\gamma}=\frac{p_2}{p_1},$$

$$p_1V_1^\gamma=p_2V_2^\gamma. \qquad \text{(15a, b, c, d, e)}$$

Equation (15e) is the equation for the adiabatic expansion of an ideal gas. You have probably seen it before.

ADIABATIC WORK - IDEAL GAS

We can use Equation (15e) as the equation for an adiabatic path on a pV diagram.

$$p^{V\gamma}=p_1V_1^\gamma$$

$$p=\frac{p_1V_1^\gamma}{V^\gamma}$$

$$p=\frac{Cons\tan t}{V^\gamma}, \qquad \text{(16a, b, c)}$$

where p_1 and V_1 refer to some arbitrary constant point on the path. Equation (16b) gives p as a function of V along the adiabatic line. We have added Equation (16c) just to emphasize the point that p_1 and V_1 refer to some fixed (constant) point on the adiabatic expansion curve. With this expression for p the work can be easily calculated,

$$dw = -pdV$$

$$w = \int_{V_1}^{V_2} pdV = -\int_{V_1}^{V_2} \frac{constant}{V^{\gamma}} dV$$

$$w = -\frac{constant}{-\gamma+1}\left(\frac{1}{V_2^{\gamma-1}} - \frac{1}{V_1^{\gamma-1}}\right) \qquad (17a, b, c, d)$$

The constant is easily found from the knowledge of one point on the adiabatic line (path).

THE JOULE-THOMPSON EXPANSION

It soon became apparent that the result of the Joule expansion experiment was not valid for real gases. A more accurate experiment, slightly different, was carried out by Joule and J. J. Thompson to further elucidate the properties on real gases under expansion.

A sample of a gas, initially at p_1, V_1, and T_1 was forced through a porous plug at constant pressure, p_1. The gas came out of the other side of the plug at p_2, V_2, and T_2. The apparatus was insulated so that $q = 0$. The work has two terms, the work done on the system to force the gas through the plug and the work done by the system on the surroundings as it came out the other side of the plug.

The total work is

$$w = p_1(0 - V_1) - P_2(V_2 - 0)$$

$$= p_1V_1 - p_2V_2 \qquad (18a, b)$$

Since $q = 0$, the change in internal energy of the gas is,

$$\Delta U = q + w$$

$$= 0 + p_1V_1 - p_2V_2$$

$$\neq 0 \qquad \text{(19a, b, c)}$$

This process, unlike the Joule expansion, is not at constant internal energy.

The enthalpy, however, is given by,

$$\Delta H = \Delta U + \Delta(pV)$$

$$= p_1V_1 - p_2V_2 + p_2V_2 - p_1V_1$$

$$\Delta H = 0 \qquad \text{(20a, b, c)}$$

So the Joule Thompson experiment is a process at constant enthalpy. In the experiment they could select a value for Δp, and then measure ΔT. The ratio of these two quantities is an approximation to a derivative,

$$\frac{\Delta T}{\Delta p} \approx \left(\frac{\partial T}{\partial p}\right)_H = \mu_{JT}. \qquad (21)$$

μ_{JT} is called the "coefficient of the Joule-Thompson effect." This coefficient is not zero for a real gas (or for realistic equations of state like the van der Waals equation of state), but we will now show that it *is* zero for an ideal gas. Applying the Euler chain rule to Equation (21) we obtain,

$$\left(\frac{\partial T}{\partial p}\right)_H = -\frac{\left(\frac{\partial H}{\partial p}\right)_T}{\left(\frac{\partial H}{\partial T}\right)_p}$$

$$= -\frac{\left(\frac{\partial H}{\partial p}\right)_T}{C_p} \qquad \text{(22a, b)}$$

The numerator in Equation (22b) is zero for an ideal gas, but not necessarily zero for a real gas.

The coefficient of the Joule-Thompson effect is important in the liquefaction of gases because it tells whether a gas cools or heats on expansion. It turns out that this coefficient is a decreasing function of temperature and it passes through zero at the Joule-Thompson inversion temperature, T_I. In an expansion $dp < 0$. Whether dT is positive or negative depends on the sign of μ_{JT}. Looking at the definition of μ_{JT},

$$\mu_{JT} = \left(\frac{\partial T}{\partial p}\right)_H,$$

we see that if μ_{JT} is positive then dT *is negative upon expansion so that the gas cools. On the other hand, if* μ_{JT} is negative, then dT is positive so that the gas warms upon expansion. In order to liquefy a gas by a Joule-Thompson expansion the gas must first be cooled to below the J-T inversion temperature. Some inversion temperatures are:

He	40 K
N_2	621 K
O_2	764 K
Ne	231 K

We see that N_2 and O_2 will cool upon expansion at room temperature, but He and Ne will warm upon expansion at room temperature.

THERMODYNAMIC EQUATION OF STATE

We have seen that the results of the Joule expansion (valid for ideal gases) demonstrated experimentally that for an ideal gas,

$$\left(\frac{\partial U}{\partial V}\right)_T = 0. \tag{1}$$

It would be advantageous to be able to calculate this quantity from an equation of state or other pVT data. There is an equation which we will prove later, but which we introduce now because it is so useful, called the "thermodynamic equation of state, which will allow us to do this. It allows us to calculate the derivative in Equation (1) from an equation of state.

The equation is,

$$\left(\frac{\partial U}{\partial V}\right)_T = T\left(\frac{\partial p}{\partial p}\right)_V - P \tag{2}$$

This equation will be proved easily once we have the second law of thermodynamics. For now we will just accept it conditionally until it can be proved. Notice that the right- hand side contains nothing but *pVT* data. We can see that the equation is at least plausible by checking that it does give zero for an ideal gas.

For an ideal gas,

$$p = \frac{nRT}{V}$$

so

$$\left(\frac{\partial p}{\partial p}\right)_V = \frac{nR}{V}.$$

Then,

$$\left(\frac{\partial U}{\partial V}\right)_T = T\left(\frac{\partial p}{\partial p}\right)_V - P$$

$$= T\left(\frac{nR}{V}\right) - p$$

$$= p - p$$

$$= 0$$

We can also check to see what our thermodynamic equation of state would give for a van der Waals gas. For the van der Waals gas we find,

$$p = \frac{nRT}{V - nb} - \frac{an^2}{V^2},$$

so

$$\left(\frac{\partial p}{\partial T}\right)_V = \frac{nR}{V - nb},$$

and

$$\left(\frac{\partial U}{\partial V}\right)_T = \left(\frac{nR}{V - nb},\right)\left(\frac{nRT}{V - nb} - \frac{an^2}{V^2}\right)$$

$$= + \frac{an^2}{V^2}.$$

We know that a is small and n^2/V^2 will be small except at very high pressures (densities).

(The above result can be understood based on what is going on in the gas. When a gas expands at constant temperature it absorbs heat from the surroundings and does work on the surroundings. If the gas is ideal the heat and work exactly balance so that there is no change in the internal energy of the gas. In a van der Waals gas - and real gases - the expansion must also overcome the intermolecular forces so part of the heat absorbed from the surroundings goes to overcoming the intermolecular forces. The a term in the van der Waals equation of state accounts for intermolecular forces. If you calculate the work in expanding a van der Waals gas you will see that that the part of the work that is proportional to a is positive so that this work was done on the system - it raised the internal energy of the system.)

In most cases

$$\left(\frac{\partial U}{\partial V}\right)_T \cdot$$

is still pretty small, even for a van der Waals gas.

There is a companion to Equation (2),

$$\left(\frac{\partial H}{\partial p}\right)_T = V - T\left(\frac{\partial V}{\partial T}\right)_p. \tag{3}$$

This equation can be derived (without the second law) from Equation (2) so that if Equation (2) is correct, so is Equation (3).

We will leave it to the reader to show that Equation (3) gives zero for an ideal gas. Applying this equation to the van der Waals gas is a little more involved and not particularly enlightening.

RELATIONSHIP BETWEEN C_p AND C_V

C_p and C_V are related to each other and their difference can be calculated from an equation of state. We wish to prove that

$$Cp = C_V + T\left(\frac{\partial p}{\partial T}\right)_V \left(\frac{\partial V}{\partial T}\right)_p . \tag{8d}$$

Let's begin with the definitions of C_p and H,

$$Cp = \left(\frac{\partial H}{\partial T}\right)_p = \left(\frac{\partial [U + pV]}{\partial T}\right)_p$$

$$= \left(\frac{\partial U}{\partial T}\right)_p + p\left(\frac{\partial V}{\partial T}\right)_p . \tag{4a, b, c}$$

The second term in (4c) is in an acceptable form, but the first term is not. (The wrong variable is being held constant.) To deal with the first term regard U as $U = U(T,V)$. Then,

$$dU = \left(\frac{\partial U}{\partial T}\right)_V dT + \left(\frac{\partial U}{\partial V}\right)_T dV . \tag{5}$$

Now divide Equation (5) by dT and hold p constant. (Your calculus teacher won't like this, but you can prove that the result is correct and that this procedure will always work.) We obtain,

$$\left(\frac{\partial U}{\partial T}\right)_p = \left(\frac{\partial U}{\partial T}\right)_V + \left(\frac{\partial U}{\partial V}\right)_T \left(\frac{\partial V}{\partial T}\right)_P . \tag{6}$$

Now we substitute Equation (6) for the appropriate term in Equation (4c) to get,

$$C_p = \left(\frac{\partial U}{\partial T}\right)_p + p\left(\frac{\partial V}{\partial T}\right)_p$$

$$= \left(\frac{\partial U}{\partial T}\right)_V + \left(\frac{\partial U}{\partial V}\right)_T \left(\frac{\partial V}{\partial T}\right)_p + \left(\frac{\partial V}{\partial T}\right)_p . \qquad (7,\text{a b})$$

Bút

$$\left(\frac{\partial U}{\partial V}\right)_T = T\left(\frac{\partial p}{\partial V}\right)_V - P. \qquad (2)$$

Substituting Equation (2) in for $(\partial U/\partial T)_V$ in Equation (7b) gives

$$C_p = \left(\frac{\partial U}{\partial T}\right)_V + \left(\frac{\partial U}{\partial V}\right)_T \left(\frac{\partial V}{\partial T}\right)_p + p\left(\frac{\partial V}{\partial T}\right)_p$$

$$= \left(\frac{\partial U}{\partial T}\right)_V + \left(T\left(\frac{\partial p}{\partial T}\right)_V - p\right)\left(\frac{\partial V}{\partial T}\right)_p + p\left(\frac{\partial V}{\partial T}\right)_p$$

$$= C_V + T\left(\frac{\partial p}{\partial T}\right)_V \left(\frac{\partial p}{\partial T}\right)_p - p\left(\frac{\partial V}{\partial T}\right)_p + \left(\frac{\partial V}{\partial T}\right)_p$$

$$= C_V + T\left(\frac{\partial p}{\partial T}\right)_V \left(\frac{\partial p}{\partial T}\right)_p , \qquad (8\text{a, b, c, d})$$

Which is the desired result. We will let the reader show, using Euler's chain relation and the definitions of α and k, that this relation can be rewritten as

$$C_p = C_V + TV\frac{\alpha^2}{k}. \qquad (9)$$

The second term on the right of Equation (9) is necessarily positive because k is always positive. α can be negative (water near 0ºC), but it appears here as the square. Thus $C_p > C_V$. For solids and liquids the second term on the right of Equation (9) is usually small. For gases it can be large. For an ideal gas we found earlier that $\alpha = 1/T$ and $k = 1/p$ so that

Chapter 5

Thermochemistry

Thermochemistry is the subject that deals with the heats involved in chemical reactions. A typical chemical reaction might have a form similar to the following hypothetical chemical reaction:

$$a\,\text{A} + b\,\text{B} \rightarrow c\,\text{C} + d\,\text{D}. \tag{1}$$

(In Equation (1) the upper case letters stand for elements or compounds and the lower case letters stand for small whole numbers which balance the reaction. You would read this as saying, "*a* moles of A reacts with *b* moles of B to give *c* moles of C and *d* moles of D.")

A chemical reaction is a process just like any other thermodynamic process. It has an initial state (the reactants) and a final state (the products). We can calculate the changes in internal energy, enthalpy, and so on for the reaction. For example,

$$\Delta U = U_{\text{products}}'' \; U_{\text{reactants}}$$

and

$$\Delta H = H_{\text{products}}'' \; H_{\text{reactants}}$$

One thing is sometimes not made very clear. "Reactants" and "products" in these equations means that the reactants and products are separated, isolated, and pure. Furthermore, the reactants and products are all at the same temperature and pressure. So, for example, the ΔH above is the enthalpy of c

moles of C (isolated and pure in its own container at temperature, T, and pressure, p) plus the enthalpy of d moles of D (isolated and pure in its own container at temperature, T, and pressure, p) minus the enthalpy of a moles of A (isolated and pure in its own container at temperature, T, and pressure, p) minus the enthalpy of b moles of B (isolated and pure in its own container at temperature, T, and pressure, p).

From time to time we will add a superscript ^{o} to H or U to indicate that reactants and products are in their "standard states." That is, they are in their most stable state at T and p. For example, the standard state of water at 25°C and 1 atm pressure is liquid water.

If our reaction takes place at constant V, as in a bomb calorimeter, $dV = 0$ and

$$\Delta U_V = q_V.$$

If the reaction takes place at constant p, as in open to atmospheric pressure, $dp = 0$ and

$$\Delta H_p = q_p.$$

If $\Delta H_p < 0$ we say that the reaction is exothermic. That is, the system gave heat to the surroundings. On the other hand, if $\Delta H_p > 0$ we say that the reaction is endothermic. The system absorbed heat from the surroundings.

From the definition of enthalpy we find that

$$\mathrm{D}H = \mathrm{D}U + \Delta(pV), \tag{2}$$

where

$$\Delta\,(pV) = (pV)_{\text{products}} - (pV)_{\text{reactants}}.$$

For liquids and solids $\Delta(pV)$ is quite small. $\Delta(pV)$ is not necessarily small for gases, but we can get a reasonable estimate for this quantity by approximating the gases as ideal. Then

$$\Delta\,(pV)_{\text{gas}} \approx (pV)_{\text{gas products}} - (pV)_{\text{gas reactants}} \approx n_{\text{gas products}} RT - n_{\text{gas reactants}} RT = RT\Delta\, n_{\text{gas}},$$

where Δn_{gas} is the difference in the number of moles of gaseous products and reactants. Using this approximation we get

$$DH = DU + RT\Delta n_{gas}. \quad (3)$$

However, we have to be careful how we understand this equation because the conditions of the reaction must be the same on both sides of the equation. Since ΔH is the heat we measure if the reaction is run and constant pressure ($\Delta H_p = q_p$) and ΔU is the heat we measure if the reaction is run at constant volume ($\Delta U_V = q_V$), it is tempting (and common) to write Equation (3) as

$$q_p = q_V + RT\Delta n_{gas}. \quad (4)$$

However, Equation (4) cannot be rigorously true since the q's refer to different conditions, one at constant p and one at constant V. We can get an indication whether or not Equation (4) is a good approximation with the following example:

Consider the reaction.

$$2\,C(s) + O_2(g) \rightarrow 2\,CO(g), \quad (5)$$

run at $dV = 0$.

$$\Delta(pV) \approx \Delta(pV)_{gas} \approx RT\Delta n_{gas} = RT.$$

The heat measured is q_V. Now let us find a way to measure q_p Consider the following two

steps

$$2\,C(s) + O_2(g) \rightarrow 2\,CO(g) \rightarrow 2\,CO(g) \quad (6)$$

$$p_1, V_1, T, q_V \quad p_2, V_1, T, \Delta H_2 \quad p_1, V_2, T.$$

The first step is the constant volume reaction we had before with $\Delta U_V = q_V$. Notice that the pressure increases. The second step takes the product of the constant volume reaction and reduces the pressure back to the original pressure. We call the heat for this step ΔH_2. So it is rigorously true that

$$\Delta H_p = \Delta H_V + \Delta H_2 = q_V + RT + \Delta H_2. \quad (7)$$

However, the ΔH_2 term is the enthalpy for the expansion of a gas at constant temperature. If the gas is ideal this term is zero. For real gases this term would be very small, so we make a negligible error is neglecting it. So to pretty good approximation we can use the equation,

$$q_p = q_V + RT\Delta n_{gas}. \qquad (4)$$

HESS' LAW

Hess' law states that if you add or subtract chemical reaction equations you can (must) add or subtract their corresponding ΔH's or ΔU's to get ΔH or ΔU for the overall reaction. For example, if we add the reaction

$$a\,\mathrm{A} + b\,\mathrm{B} \rightarrow c\,\mathrm{C} + d\,\mathrm{D}. \qquad \Delta H_1, \Delta U_1$$

to the reaction

$$e\,\mathrm{E} + f\,\mathrm{F} \rightarrow g\,\mathrm{G} + h\,\mathrm{H} \qquad \Delta H_2, \Delta U_2$$

to get

$$\mathrm{A} + b\,\mathrm{B} + e\,\mathrm{E} + f\,\mathrm{F} \rightarrow c\,\mathrm{C} + d\,\mathrm{D} + g\,\mathrm{G} + h\,\mathrm{H},$$

Then, for the overall reaction

$$\Delta H = \Delta H_1 + \Delta H_2 \text{ and } \Delta U = \Delta U_1 + \Delta U_2.$$

The great utility of Hess' law is that we don't have to tabulate ΔH for every possible reaction. We can get ΔH for a particular reaction by adding and subtracting ΔH's for a much smaller set of reactions, called formation reactions. We define $\Delta_f H^\circ$ for a compound to be the enthalpy of the reaction:

pure, isolated elements, in their standard (most stable) states

$\rightarrow$one mole of compound in its standard state.

For example, the heat of formation of liquid water is defined as ΔH° for the reaction,

$$H_2(g) + 1/2\ O_2(g) \rightarrow H_2O(l).$$

By definition, then, the heat of formation for an element in its most stable state is zero.

To obtain ΔH° for the hypothetical reaction, Equation (1), we add and subtract the appropriate heats of formations,

$$\Delta_r H^\circ = c\Delta_f H_C{}^\circ + d\Delta_f H_D{}^\circ - a\Delta_f H_A{}^\circ - b\Delta_f H_B{}^\circ \qquad (8)$$

For example, ΔH° for the reaction,

$$CH_4(g) + 2\,O_2(g) \rightarrow CO_2(g) + 2\,H_2O(l), \qquad (9)$$

is given by

$$\Delta_r H^\circ = \Delta_f H^\circ{}_{CO_2(g)} + 2\Delta_f H^\circ{}_{H_2O(l)} - \Delta_f H^\circ{}_{CH_4(g)}$$

$$1\,\text{mol}\times(-393.51\frac{\text{kJ}}{\text{mol}}) + 2\,\text{mol}\times(-285.83\frac{\text{kJ}}{\text{mol}}) - 1\,\text{mol}\times(-74.81\frac{\text{kJ}}{\text{mol}})$$

$$= -890.36\,\text{kJ}.$$

We don't always write out explicitly that the coefficients which balance the reaction have units, but we have done so here to make it clear that these numbers have units. This will become an issue later.

Δ *H* AT OTHER TEMPERATURES

Tables of heats of formation usually give data for reactions at 25°C. We frequently need to know the heat of a reaction at a temperature other than 25°C. If we know the heat capacities at constant pressure we can calculate the heat of reaction at a temperature other than 25°C. We use the following chain of reasoning. We know that,

$$\left(\frac{\partial H}{\partial T}\right)_p = C_p$$

then

$$\left(\frac{\partial H^\circ}{\partial T}\right)_p = C_p^o \text{ and}$$

$$\left(\frac{\partial \Delta H^\circ}{\partial T}\right)_p = \Delta C_p^o, \qquad (10a, b, c)$$

where $\Delta C_p{}^o$ is defined for our hypothetical chemical reaction, Equation (1) as,

$$\Delta C_p^o = cC^\circ{}_{pc} + dC^\circ{}_{pD} - aC^\circ{}_{PA} - bC^\circ{}_{PB} \tag{11}$$

Prepare Equation 10c for integration as

$$d\Delta H^\circ = \Delta C_p^o\, dT, \tag{12}$$

and integrate,

$$\Delta H^\circ(T_2) = \Delta H^\circ(T_1) + \int_{T_1}^{T_2} \Delta C_p^o dT. \tag{13}$$

For very accurate work we will have to use the temperature dependent heat capacities in Equations (11) and (13), but more often than not we can regard the heat capacities as approximately constant over the temperature range, so that Equation (13) becomes,

$$\Delta H^\circ(T_2) = \Delta H^\circ(T_1) + \Delta C_p^o(T_2 - T_1). \tag{14}$$

As an example, let's calculate the heat of reaction for the reaction in Equation (9) at 95°C.

$\Delta C_p{}^o$ for the reaction in Equation (9) is given by,

$$\Delta_r C_p^o = C^o_{pCO_2(g)} + 2C^o_{pH_2O(l)} - C^o_{pCH_4(g)} - 2C^o_{pO_2(g)}$$

$$1mol \times (37.11\frac{J}{Kmol}) + 2mol \times (75.291\frac{J}{Kmol})$$

$$-1mol \times (35.35\frac{J}{Kmol}) - 2mol \times (29.355\frac{J}{Kmol})$$

$$= 93.67\frac{J}{K}.$$

Then, Equation (14) gives,

$$\Delta H^\circ = (95°C) = \Delta H^\circ(25°C) + \Delta C_p^o(368.15 - 298.15)$$

$$= -890.36kJ + 93.67\frac{J}{K} \times 70K \frac{1kJ}{1000J}$$

$$= \text{-883.8. kJ} \qquad (16)$$

(Note: There is another way to do this problem. Calculate ΔH^{o} for cooling the reactants down to 25°C, calculate ΔH^{o} for the reaction at 25°C, calculate ΔH^{o} for heating the products back up to 95°C, and then add them up. Since H is a state function ΔH is independent of path. This method will also work if one of the components of the reaction has a phase change somewhere in the temperature range.

For example, if we let the new temperature be over 100°C we would have to account for the vapourization of the liquid water product. This is not hard to do, but requires some extra steps, including also the use of the heat capacity of water vapour. If the upper temperature in this problem is above 100°C we carry out the reaction in several steps:

Step 1 - Cool the reactants from the upper temperature to 25°C,

Step 2 - Run the reaction at 25°C,

Step 3 - Heat the product CO_2 from 25°C to the upper temperature,

Step 4 - Heat the liquid water from 25°C to 100°C,

Step 5 - Vaporize the water at 100°C,

Step 6 - Heat the water vapour from 100°C to the upper temperature.

The heat of reaction at the upper temperature is the sum of the ΔH's for all six steps. Again we have taken advantage of the fact that ΔH is independent of path.)

ΔH AS MAKING AND BREAKING CHEMICAL BONDS

Breaking a chemical bond is an endothermic process. That is, you must put energy into the system to break the bond.

Forming a chemical bond is an exothermic process. The energy released in forming the bond goes into the surroundings.

We can make an estimate of the ΔH for a chemical reaction by adding the bond energies of all the bonds broken and subtracting the bond energies of all the bonds formed.

$$\Delta H \approx BE_{\text{bonds broken}} \text{''} BE_{\text{bonds formed}}.$$

Let's try it on the gas phase reaction,

$N_2 + 3\ H_2 \rightarrow 2\ NH_3$.

From the tables we find the following bond energies:

N–N 945 kJ

H–H 436 kJ

N–H 388 kJ.

In the reaction we break 1 N–N bond and 3 H–H bonds. We form 6 N–H bonds. The approximate ΔH is then,

ΔH H $\approx$ 1 × 945 + 3 × 436 " 6 × 388 = –75 kJ.

This can be compared to the actual value of " 92 kJ. The method is not super accurate, but it gives a ball-park answer and might be useful in cases where other data are not available. Further, however it demonstrates graphically that the heat of a reaction is related to the making and breaking of chemical bonds.

HEATS OF FORMATION OF IONS IN WATER SOLUTION

When you look in a table of heats of formation you find values listed for ions in water solutions. That is, you will find an entry for species such as $Na^+(aq)$. It is fair to ask where these numbers come from. We know that in equilibrium chemistry it is impossible to prepare $Na^+(aq)$ ions in solution all by themselves. It is possible to prepare a solution that has both $Na^+(aq)$ and $Cl^-(aq)$ ions, but not a solution that has only ions of one charge.

The heats of formation of solutions of soluble ionic compounds can be measured. That is, we can measure the heat of formation of HCl(*aq*). The heat of formation of HCl(*aq*) is defined as ΔH° for the reaction,

$1/2\ H_2(g) + 1/2\ Cl_2(g) \rightarrow HCl(aq)$. (17)

Since ionic compounds in solution are completely dissociated, it must be true that

$$\Delta_f H^{\circ}{}_{HCl(aq)} = \Delta_f H^{\circ}{}_{H^+(aq)} + \Delta_f H^{\circ}{}_{Cl^-(aq)}. \qquad (18)$$

We cannot know the heat of formation of either of the ions in solution, but we do know their sum.

By convention we arbitrarily set the heat of formation of the $H^+(aq)$ ion equal to zero. That is,

$$\Delta_f H^{\circ}{}_{H^+(aq)} = 0 \qquad (19)$$

Then the heats of formation of all other aqueous ions can be determined relative to the heat of formation of the $H^+(aq)$ ion. As a start, we see that

$$\Delta_f H^{\circ}{}_{Cl^-(aq)} = \Delta_f H^{\circ}{}_{HCl(aq)} - \Delta_f H^{\circ}{}_{H^+(aq)}$$

$$\Delta_f H^{\circ}{}_{HCl(aq)} - 0$$

$$\Delta_f H^{\circ}{}_{HCl(aq)} \qquad (20a, b, c)$$

This convention allows us to build up a table of heats of formation of aqueous ions.

For example, from the measured value of the heat of formation of NaCl(*aq*) and the knowledge that

$$\Delta_f H^{\circ}{}_{NaCl(aq)} = \Delta_f H^{\circ}{}_{Na^+(aq)} + \Delta_f H^{\circ}{}_{Cl^-(aq)} \qquad (21)$$

we find that

$$\Delta_f H^{\circ}{}_{Na^+(aq)} = \Delta_f H^{\circ}{}_{NaCl(aq)} - \Delta_f H^{\circ}{}_{Cl^-(aq)}$$

$$= \Delta_f H°_{Nacl(aq)} + \Delta_f H°_{HCl(aq)} \quad (22a, b)$$

Continuing these procedures we can define the heats of formations of other aqueous ions,

$$\Delta_f H°_{Br^-(aq)} = \Delta_f H°_{HBr(aq)} + \Delta_f H°_{H^+(aq)}$$

$$\Delta_f H°_{HBr(aq)} - 0$$

$$\Delta_f H°_{HBr(aq)^-} \quad (23a, b, c)$$

We now have enough data in our table to calculate the heat of formation of aqueous NaBr without measuring it.,

$$\Delta_f H°_{NaBr(aq)} = \Delta_f H°_{Na^+(aq)} + \Delta_f H°_{Br^-(aq)}$$

$$= \Delta_f H°_{Nacl(aq)} + \Delta_f H°_{Hcl(aq)} + \Delta_f H°_{HBr(aq)^-} \quad (24a, b)$$

If the heat of formation of $H^+(aq)$ were set to some value other than zero, it would have cancelled out of Equation (24b). In this manner an entire table of heats of formation of aqueous ions can be built with all values relative to the heat of formation of the $H^+(aq)$ ion.

EXACT AND INEXACT DIFFERENTIALS

We have mentioned, from time to time, that the quantities, U, H, and so on, are state functions, but that q and w are not state functions. This has various consequences. One consequence is that we can write things like ΔU and ΔH, but we never write q or w with a Δ in front them. A more important consequence is that in a process ΔU and ΔH are independent of path. That is, ΔU and ΔH depend only on the initial and final states. However, q and w do depend on the path one takes to get from the initial to the final state.

Another consequence is that the differentials, dU and dH are mathematically different, in some sense, from dq and dw. Some writers write dq and dw with a line through the d to indicate this difference. We have not chosen to use such a specialized notation, but expect that we all will be able to just remember that dU and dH are mathematically different, in some sense, than dq and dw.

We must now consider in detail the nature of this difference.

Let's think, for the moment, in terms of functions of the variables x and y and consider the differential,

$$df = 2xy^3dx + 3x^2y^2dy. \tag{1}$$

We ask the question, does there exist a function, $f = f(x,y)$ such that,

$$df = \left(\frac{\partial f}{\partial x}\right)_y dx + \left(\frac{\partial f}{\partial x}\right)_x dy. \tag{2}$$

In other words, does a function $f(x,y)$ exist such that,

$$\left(\frac{\partial f}{\partial x}\right)_y = 2xy^3$$

and

$$\left(\frac{\partial f}{\partial x}\right)_y = 3x^2y^2\,? \tag{3a, b}$$

Euler's test provides a way to see whether such a function, $f(x,y)$ exists. Euler's test is based on the fact that for "nice" functions (and all of our functions are "nice") the mixed second derivatives must be equal. That is,

$$\left(\frac{\partial^2 f}{\partial x \partial y}\right) = \left(\frac{\partial^2 f}{\partial y \partial x}\right). \tag{4}$$

(On the left-hand side we take the derivative with respect to y first and then take the derivative of the result with respect to x, and vice versa on the right-hand side.

Let's try Euler's test on our differential, df. If f exists then the Equations (3a) and (3b) are correct. Use Equations (3a, b) to obtain the proposed second derivatives,

$$\frac{\partial}{\partial y}\left(\frac{\partial f}{\partial x}\right)_y = \frac{\partial}{\partial y}2xy^3 = 6xy^2$$

and

$$\frac{\partial}{\partial x}\left(\frac{\partial f}{\partial y}\right)_x = \frac{\partial}{\partial x}3x^2y^2 = 6xy^2 \qquad \text{(5a, b)}$$

The mixed second derivatives are equal. So we conclude that there exists a function, $f(x,y)$ (actually x^2y^3) such that Equations (1) and (2) are equal.

The differential df, in Equation (1), is called an "exact differential" for the very reason that a function, f, exists such that Equation (2) can be used to calculate it.

Now, let's consider the differential,

$$dg = 2x^2y^3dx + 3x^3y^2dy. \qquad (6)$$

Is dg an exact differential? Use Euler's test to find out. If dg is exact then the coefficients of dx and dy are the respective partial derivatives of g. Euler's test would then compare

$$\frac{\partial}{\partial y}2x^2y^3 = 6x^2y^2$$

$$\frac{\partial}{\partial y}3x^3y^2 = 9xy^2. \qquad (7)$$

These are not equal so that the putative second partial derivatives are not equal to each other. The differential, dg, is not exact and there does not exist a function, $g(x,y)$, such that dg gives Equation (6).

Both of the differentials, df and dg can be integrated from, say, x_1, y_1 to x_2, y_2. The integral,

$$\int df = f(x_2, y_2) - f(x_1, y_1). \qquad (8)$$

depends only on the initial and final points because df is exact and the function f exists.

The differential dg can be integrated, but there is no equivalent to Equation (8) for the integral of dg because there is no function, $g(x,y)$ which gives Equation (6). The integral of dg would have to be carried out along some path and we would find that the value of the integral depends on the path as well as on the initial and final points.

So, what is the purpose of all this? We are getting ready to present the second law of thermodynamics. One of the consequences of the second law will be the demonstration that for a reversible process dq/T is exact. dq_{rev} is not exact, but dq_{rev}/T is exact. That means that dq_{rev}/T is the differential of some new function (a state function) whose integral is independent of path. We will call the new state function, S, and name it the "entropy."

The absolute temperature, T, is called an "integrating denominator" for dq_{rev}. That is, when we divide the inexact differential, dq_{rev}, by T the resulting differential becomes exact. Notice that the inexact differential, dg above, has an integrating denominator. The variable, x, is an integrating denominator for dg. You can see this by noticing that $dg/x = df$.

SECOND LAW OF THERMODYNAMICS

When we introduced the first law of thermodynamics we claimed that it is a statement of repeated observation elevated to the status of a law. No one has ever been able to make a machine that produces work out of nothing (a perpetual motion machine of the first kind), so we assume that no such machine can be made.

We then write this statement in mathematical language and begin deriving the consequences of the statement. Ultimately, the validation of the law comes from the experimental verification of the consequences.

The second law of thermodynamics is also a statement of repeated observation (or perhaps better yet, a statement of some things that have never been observed).

Here are two things that have never been observed:

1. Heat has never been observed to move spontaneously from a cold body to a hot body.
2. Heat has never been observed to be converted entirely into work with no other result.

So the second law, in words, is just the statement that these two things are impossible. that is:

1. It is impossible for heat to move spontaneously from a cold body to a hot body with no other result.
2. It is impossible to convert heat quantitatively into work with no other result.

The latter statement is sometimes phrased: "It is impossible to make a perpetual motion machine of the second kind."

(A perpetual motion machine of the second kind is a machine that converts heat into work without doing anything else. Imagine an ocean liner that scoops up liquid water out of the ocean, pulls the heat out of the water and uses it to power the ship, and dumps the left-over ice cubes out the back of the ship.

Note that a perpetual motion machine of the second kind would not violate the first law. Energy would be conserved because any heat extracted would be converted into work.)

The second law is why automobiles have radiators. Someone might ask why we throw away all that energy that dissipates from the radiator. Why not capture the energy and use it do decrease our gas mileage? The answer is that if you don't dissipate the heat the engine burns up, as you would quickly find out if you bypassed the radiator with a hose or if you drained the coolant from the radiator.

In order to convert these word statements into mathematical statements we can use we will have to develop some apparatus.

First we define the "heat engine." A heat engine is a cyclic process that absorbs heat from a heat bath and converts it into work. We shall see that in the cyclic process the engine also dissipates some heat to a heat bath at a lower temperature.

A crucial feature of the heat engine is that it returns to its original state after each cycle. That means that for each cycle of the engine itself, $\Delta H = 0$, $\Delta U = 0$, $\Delta T = 0$, and so on. Presumably, less heat is given back at the lower temperature than was absorbed at the upper temperature so that the difference can be used to supply work to the surroundings. (Otherwise we wouldn't have much of an engine.)

If you run the engine backwards by providing an external power source you get a heat pump (or a refrigerator), that is, a machine that absorbs heat from a lower temperature heat bath and gives it back to a heat bath at a higher temperature. But it takes work to do this.

Our procedure will be as follows:

1. Define and characterize a particular heat engine, the Carnot Cycle. The Carnot cycle is a heat engine operating between two heat baths, one at an upper temperature, which we shall call T_U and the other at a lower temperature, T_L. The Carnot cycle uses the expansion and compression of an ideal gas to convert heat into work.
2. Define the efficiency, *e*, of a Carnot cycle.
3. Assume that we can find a heat engine, operating between the same two temperatures, which has efficiency greater than a Carnot cycle efficiency and then show that this violates both of the word statements of the second law given above. This leads to the conclusion that no heat engine or cycle can have an efficiency greater than the efficiency of a Carnot cycle.
4. The conclusion that no cycle can have an efficiency greater than a Carnot cycle will lead us to the further conclusion that the integral of dq_{rev}/T is independent of path. Therefore, the differential dq_{rev}/T must be

exact, which means that it is the differential of some state function which we will call, S. That is, $dq_{rev}/T = dS$.

5. Another cycle can have an efficiency *less than* the efficiency of a Carnot cycle. This will lead us to the conclusion that dq/T might be less than dS if there is some irreversibility in the process. Putting the two possibilities together we will conclude that, dq/T d" dS. This is the mathematical statement of the second law of thermodynamics.

One further comment. We have seen processes where heat is converted into work before. The isothermal reversible expansion

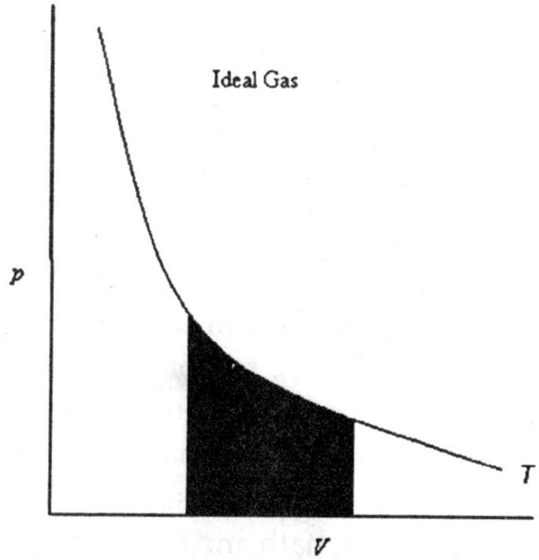

of an ideal gas is such a process. Recall that the work is the negative of the area under the curve. However, this is not a cyclic process. A change has taken place: the volume has increased and the pressure has decreased. In order to get more work out of the system you would have to expand the gas even further. This is *not* a process in which heat was converted into work and nothing else happened.

HEAT ENGINES AND THE CARNOT CYCLE

A heat engine is a cyclic process that absorbs heat and does work on the surroundings. "Cyclic" means that the system returns to its initial state at the end of each cycle so that there is no permanent change in the system. The only cycle we will work with in this course is the Carnot Cycle which is shown below on a *p*-*V* diagram.

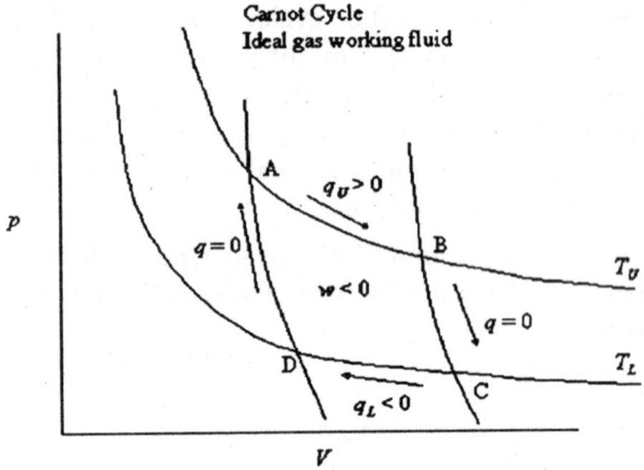

The curves labeled T_U and T_L are isotherms. T_U is the temperature of the upper temperature heat bath (reservoir) and T_L is the temperature of the lower temperature heat bath. The two steep curves, BC and DA, are adiabatic curves. The cycle begins at the point A. The system undergoes an isothermal expansion at temperature, T_U, to point B. In this isothermal expansion the system absorbs heat from the upper heat bath ($q_U > 0$) and does work on the surroundings (recall that w is defined as work done on the system so $w_{AB} < 0$). The system is then isolated from the heat bath and is expanded adiabatically to the point C. There is no heat in this adiabatic expansion, but the work for this step is also negative ($w_{BC} < 0$). At point C the system is placed in contact with a heat bath at T_L and undergoes an isothermal compression to point D. For this segment of the cycle $q_L < 0$ and $w_{CD} > 0$ because the surroundings are now doing work *on* the system and heat is being dissipated to the heat bath at T_L. At point D the system is again isolated from the heat baths and compressed adiabatically to point A. In this adiabatic compression the heat, of course, is zero and the work is positive ($w_{DA} > 0$)

We can analyze the Carnot cycle as follows: The heat, q, for the whole cycle is

$$q = q_{AB} + q_{CD},, \quad (1)$$

and the work for the entire cycle is,

$$w = w_{AB} + w_{BC} + w_{CD} + w_{DA}\,. \quad (2)$$

Since the initial state of the cycle is the same as the final state we know that the change in U, the internal energy, is zero. (That's part of the first law, because the first law says that U is a state function, which means that the value of U for any state of the system does not depend on how the system got to that state, only on what the state actually is.) The first law of thermodynamics then tells us that

$$\Delta U = q + w = 0, \quad (3)$$

from which we see that

$$q = -w \quad (4)$$

Notice that when the Carnot cycle is operating as a heat engine (going around clockwise), $w < 0$, so that $-w > 0$ and $q > 0$.

As we have said, the Carnot cycle is a heat engine. That means that we want it to absorb heat and convert that heat energy into work. We certainly should care about how much work we get out for the heat we absorb. Thus we will define the efficiency, e, of a cycle as

$$e = \frac{-w}{q_{AB}} \quad (5)$$

which, when combined with Equations 1 and 4 gives

$$e = \frac{q_{AB} + q_{CD}}{q_{AB}} \quad (6)$$

In order to conform to the usual notation we note that the $A \rightarrow B$ segment of the cycle is at the temperature of the upper heat bath, T_U, and the segment C → D is at the temperature of the lower heat bath, T_L. Thus equation 6 becomes

$$e = \frac{q_U + q_L}{q_U} = 1 + \frac{q_L}{q_U} \tag{7}$$

The A → B and C → D segments are an isothermal expansion of an ideal gas at T_U and an isothermal compression at T_L, respectively. For isothermal expansions and compressions of ideal gases, . Thus we can write

$$q_U = -w_{AB} = +nRT_U In \frac{V_B}{V_A} \tag{8}$$

and

$$q_L = -w_{CD} = +nRT_L In \frac{V_D}{V_C} \tag{9}$$

Let's plug these heats into Equation 7, (right hand side) to get,

$$e = 1 + \frac{nRT_L In \frac{V_D}{V_C}}{nRT_U In \frac{V_B}{V_A}} \tag{10}$$

The nR cancels leaving,

$$e = 1 + \frac{T_L In \frac{V_D}{V_C}}{T_U In \frac{V_B}{V_A}} \tag{11}$$

If we are looking for an expression for efficiency in terms of temperatures Equation 11 won't do the job because of all the volumes in the equation. However, there is another set of relations between the volumes from the fact that the paths B → C and D → A are adiabatic and we know the volume relationships for adiabatic processes in an ideal gas. Recall that for an adiabatic expansion of an ideal gas we had the expression (using the temperatures, pressures, and volumes appropriate to our Carnot cycle here)

$$p_B V^{\gamma} B = p_C V \gamma c \tag{12}$$

for the B$\rightarrow$C segment of the cycle and a similar equation for the D$\rightarrow$A segment. (Recall that γ is the "heat capacity ratio," C_p/C_V.) Let us eliminate the pressures from Equation 12 by inserting the value of pressure from the ideal gas equation of state, remembering that at points A and B the temperature is T_U and at points C and D the temperature is T_L.

$$\frac{nRT_U}{V_B} V^{\gamma}{}_B = \frac{nRT_L}{V_C} V^{\gamma}{}_C \tag{13}$$

Cancel nR from both sides and combine the volumes to get,

$$T_U V_B^{\gamma-1} = T_L V_C^{\gamma-1} \tag{14}$$

From which we obtain

$$\frac{T_U}{T_L} = \frac{V_C^{\gamma-1}}{V_B^{\gamma-1}} \tag{15}$$

Note that there is an equivalent expression for the D$\rightarrow$A leg of the cycle,

$$\frac{T_U}{T_L} = \frac{V_D^{\gamma-1}}{V_A^{\gamma-1}} \tag{16}$$

From these two last equations we conclude that

$$\frac{V_C^{\gamma-1}}{V_B^{\gamma-1}} = \frac{V_D^{\gamma-1}}{V_A^{\gamma-1}} \tag{17}$$

or

$$\frac{V_C}{V_B} = \frac{V_D}{V_A} \tag{18}$$

or, on rearranging,

$$\frac{V_A}{V_B} = \frac{V_D}{V_C} \tag{19}$$

We now apply this result to Equation 11 to get,

$$e = 1 + \frac{T_L In \frac{V_A}{V_B}}{T_U In \frac{V_B}{V_A}} \tag{20}$$

The two logarithms are the negatives of each other so we conclude that,

$$e = 1 - \frac{T_L}{T_U} = \frac{T_U - T_L}{T_U} \tag{21}$$

Notice that the highest efficiencies are obtained with a low temperature for the heat bath at T_L and that you could only obtain unit efficiency if the lower heat bath were at absolute zero. Since, in most cases, T_L is constrained by the surroundings engineers try to use the highest feasible operating temperature, T_U . Real heat engines, of course, are not Carnot cycles, but the second law of thermodynamics requires that no heat engine operating between T_U and T_L can have an efficiency greater than a Carnot cycle efficiency. So the Carnot cycle provides an idealized upper limit to the efficiency of heat engines. Even though the Carnot cycle is idealized the general principles of heat engines remain the same. If your lower temperature, T_L, is constrained (by design or operating considerations) you can increase the efficiency of your heat engine by increasing T_U. If your upper temperature, T_U, is constrained you can increase the efficiency by lowering T_L.

You may have noticed that all of our temperatures are ideal gas temperatures. That is,they are measured on the ideal gas temperature scale. It is possible, and desirable to use the

Carnot cycle to define a "thermodynamic temperature scale." We will not do this here. Suffice it to say that by choosing a suitable reference temperature you can make the two scales identical.

Second Law of Thermodynamics - Two Cycles

We have seen that the Carnot cycle, a reversible heat engine with an ideal gas working fluid, has an efficiency,

$$e = \frac{T_U - T_L}{T_U} \tag{1}$$

Is this the best we can do? Does there exist another cycle which has an efficiency greater than the efficiency of the Carnot cycle? We will now show that if such a cycle exists, with an efficiency greater than a Carnot cycle, then both of our word statements of the second law will be violated. That is, if a cycle (a heat engine working between the two heat reservoirs at T_U and T_L) exists with an efficiency greater than the efficiency of a Carnot cycle, then we can see heat spontaneously moving from a low temperature to a higher temperature with no other effect, *and* we will be able to convert heat quantitatively into work with no other effect.

Let us call the cycle with higher efficiency the "better" cycle and just say about it that it has an efficiency, *e′*, which is greater than *e*.

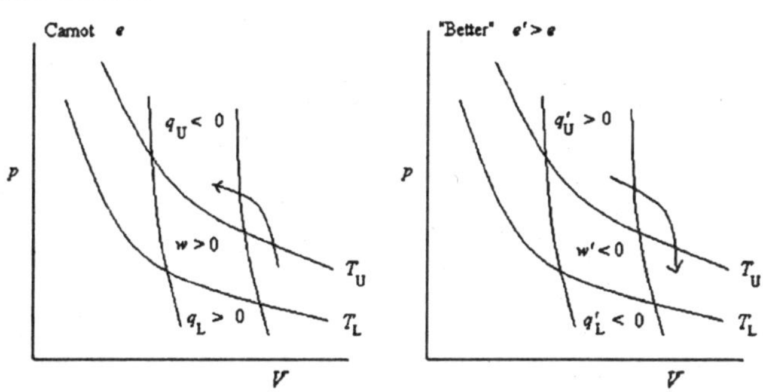

We will set up the two cycles such that the "better" cycle drives the Carnot cycle. That is, the "better" cycle will be operated as a heat engine and it will drive the Carnot cycle which then operates as a heat pump. We know that $\Delta U = 0$ for each cycle independently and for the sum of the two cycles. Then,

$$\Delta U_{overall} = w + w' + q_U + q_L + q_U' + q_L' = 0, \tag{2}$$

or

$$w + w' = -(q_U + q_L + q'_U + q'_L). \tag{3}$$

We can adjust each cycle to alter the various q's and w's by changing where the adiabatic curves intersect the isotherms. Changing the positions of the adiabatic lines changes the area enclosed by the cycle and also changes the heats absorbed and released on the isothermal lines.

As we have said, in both of our two experiments the "better" cycle runs clockwise - as an engine, and the Carnot cycle runs counterclockwise as a heat pump. The engine drives the heat pump. Some or all of the work produced by the "better" cycle is used to run the (Carnot cycle) heat pump.

Experiment 1

Adjust the parameters of the two cycles such that $w' + w = 0$. That is, the work produced by the "better" cycle is entirely used up to drive the Carnot heat pump. This also means that there is no net work either on the system or on the surroundings. Then,

$$w = {}^{-}w',$$

which we can use in the expressions for the efficiencies,

$$e' > e$$

$$\frac{-w'}{q'U} > \frac{-w}{qU}. \tag{4a, b}$$

In Equation 4b use the fact that $w =$ " w' on the left-hand-side. For the right-hand-side, use the fact that q_U is negative and move the negative side to the denominator so that both numerator and denominator are positive on the right. Then,

$$e' > e$$

$$\frac{+w'}{q'_U} > \frac{-w'}{q'_U} > \frac{-w}{q_U} = \frac{w}{-q_U}$$

$$\frac{+w}{q'_U} > \frac{w}{-q_U}. \qquad (4a, b, c)$$

In Equation (4c) w is positive so we can divide both sides by w without affecting the inequality.

$$\frac{1}{q'_U} > \frac{1}{q_U'}$$

$$q'_U < -q_U, \qquad (5,a b)$$

since both q'_U and " q_U are positive. Then,

$$q'_U + q_U < 0. \qquad (6)$$

If you were to stand back and take a look at the overall effect of this experiment, including both cycles, you would see that, according to Equation (6), heat is being released into the heat bath at the upper temperature. This is heat that must have been absorbed from the heat bath at the lower temperature because of the first law and as indicated in Equation (3). However, we also know that there is no net work being done on the two systems. The work of the "better" cycle is entirely used up driving the Carnot cycle. It appears, looking at the overall effect, that heat is spontaneously disappearing from the lower temperature heat bath and appearing in the upper temperature heat bath and there is no other effect. This violates our first word statement of the second law.

Experiment 2

This time adjust the parameters of the experiment so that heat given to the lower temperature heat bath by the "better" cycle is exactly balanced by the heat absorbed at the lower temperature by the Carnot cycle. That is, set

$$q_L + q'_L = 0 \qquad (7)$$

This time we find that

$$e' > e$$

$$\frac{q'_U + q'_L}{q'_U} > \frac{q_U + q_L}{q_U}$$

$$1 + \frac{q'_L}{q'_U} > 1 + \frac{q_L}{q_U}$$

$$\frac{q'_L}{q'_U} > \frac{q_L}{q_U}. \qquad (8a, b, c, d)$$

Invert (8d) to get

$$\frac{q'_U}{q'_L} < \frac{q_U}{q_L}$$

$$\frac{-q'_U}{-q'_L} < \frac{q_U}{q_L}. \qquad (9a,b)$$

Recall that, by construction of Experiment 2, we have from Equation (7),

$$-q'_L < q_L > 0. \qquad (10)$$

Equation (9b) becomes

$$\frac{-q'_U}{q_L} < \frac{q_U}{q_L},$$

$$-q'_U < q_U,$$

$$0 < q'_U + q_U. \qquad (11a, b, c)$$

The net effect is that heat has been absorbed at the upper temperature and work has been done on the surroundings, but there was no net heat transferred to the lower temperature heat bath. This violates the second word statement of the second law. If we regard the two cycles as one large heat engine, then that engine has no "radiator." It would be nice if such an arrangement would work, and from time to time people propose schemes which are designed to make it work, but it appears that nature does not allow it.

The initial assumption that we can find another cycle, with an efficiency, *e′*, better than the Carnot cycle efficiency, leads to a contradiction of both word statements of the second law. We are forced to conclude that no cycle can have an efficiency greater than the efficiency of a Carnot cycle. That is

$$e' \leq e = \frac{T_U - T_L}{T_U} \tag{12}$$

Equation (12) is useful and has a number of interesting and important consequences, but it does not yet provide us with the most significant result of the second law. On the next page we will use Equation (12) to define a new state function, S, called "entropy," and write the second law in a mathematical form which will give us enormous new calculating power.

THE EQUATION

We have seen that there are two word statements of the second law of thermodynamics. Both statements are just statements of universal observation. That is, no one has ever observed a violation of these statements and no one expects that a violation ever will be observed. As we did with the first law, we elevate these statements to the status of a "law" and assume that they are universally valid. Then we derive the consequences of this law, which can be checked out by experiment. So far, within the domain of its validity, no one has ever observed a violation of the second law, and its consequences are consistent with experimental observation.

Recall that the word statements of the second law are:

1. Heat does not move spontaneously from a cold body to a hot body with no other effect.
2. You can not convert heat quantitatively into work with no other effect.

The first statement is pretty obvious. Heat flows from hot bodies to cold bodies not the other way around. It would be very startling if all of the heat in your pencil spontaneously flowed into the eraser end so that the eraser melted and caught fire. The second statement is probably not so obvious unless you are an engineer, but it is one of the reasons that automobiles have radiators. In practice it means that you can't convert heat into work without dissipating some of the heat into a heat bath at a lower temperature than your heat source.

We demonstrated on the previous page that both of these statements lead to the conclusion that no heat engine can have an efficiency greater than a Carnot cycle.

(Recall that the efficiency of a cycle operating as a heat engine between two heat baths, one at an upper temperature, T_U, and the other at a lower temperature, T_L, can be written in several ways. For the Carnot cycle we can write,

$$e = \frac{-w}{q_U} = \frac{T_U - T_L}{T_U} = 1 - \frac{T_L}{T_U} = \frac{q_U + q_L}{q_U} = 1 + \frac{q_L}{q_U} \tag{1}$$

Likewise, for some other cycle, which we will indicate by putting a prime on the heats and work - still operating between T_U and T_L - we can write,

$$e = \frac{-w'}{q'_U} = \frac{T_U - T_L}{T_U} = 1 - \frac{T_L}{T_U} = \frac{q'_U + q'_L}{q'_U} = 1 + \frac{q'_L}{q'_U} \tag{2}$$

We are now considering the cycles to be operating as heat engines (as opposed to refrigerators, or heat pumps) - going around the cycle clockwise, so that w and w' are negative, q_U and q'_U are positive, and q_L and q'_L are negative. (We will need to know the signs of these quantities in order to carry out the algebra of inequality signs.)

Since we know from the second law that no cycle can have an efficiency greater than a Carnot cycle, we can write,

$$e' \leq e \tag{3}$$

(We might ask what circumstances might cause the efficiency of an engine to decrease. The most obvious answer is that irreversibility, for example, friction, would cause the efficiency to go down. If there is any irreversibility in the cycle its efficiency will be degraded.)

Using our various expressions for the efficiency, Equations 1 and 2, we can rewrite equation 3 as

$$e' = 1 + \frac{q'_L}{q'_U} \le e = 1 - \frac{T_L}{T_U} \tag{4}$$

or,

$$\frac{q'_L}{q'_U} \le \frac{T_L}{T_U} \tag{5}$$

Multiply both sides of Equation 5 by " 1, which reverses the inequality,

$$\frac{-q'_L}{q'_U} \ge \frac{T_L}{T_U} \tag{6}$$

In Equation 6 we have purposely associated the minus sign with the q'_L which makes both the numerators and denominators of both sides positive. With all quantities in this equation positive we can cross-multiply at will without worrying about the direction of the inequality. Cross-multiplying we get

$$\frac{-q'_L}{T_L} \ge \frac{-q'_U}{T_U} \tag{7}$$

or

$$0 \ge \frac{q'_U}{T_U} + \frac{-q'_L}{T_L} \tag{8}$$

The equal sign holds when the cycle is fully reversible and "greater than" sign holds if there is any irreversibility in the cycle. In the case where the cycle is fully reversibly, then,

$$\frac{q'_{U,rev}}{T_U} + \frac{-q'_{L,rev}}{T_L} = 0 \tag{9}$$

This latter equation makes it look like,

$$\sum_{cycle} \frac{q_{rev}}{T} = 0 \tag{10}$$

(Recall that two of the "legs" of the cycle were adiabatic so that there is no reversible heat on those legs.) We can generalize Equation 10 to an arbitrary closed path in p, V space by noting that we can fill an arbitrary closed path with a large number of small Carnot cycles (an infinite number in the limit of small cycles).

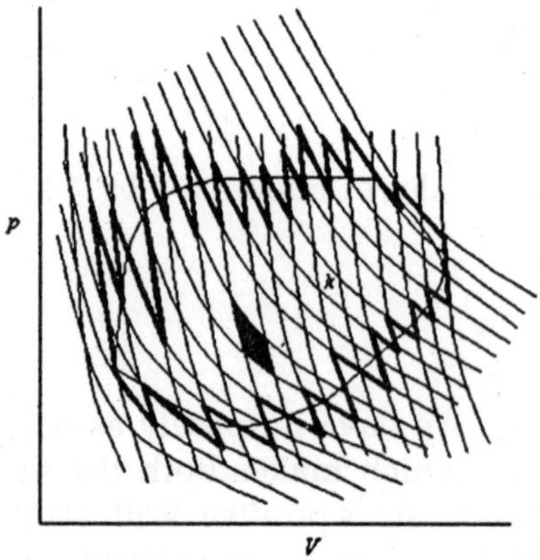

When we sum over all the cycles inside the heavy line the inside "lines" of the cycles (for example the blue filled Carnot cycle in the centre) will cancel because each inside line is traversed once clockwise and once counterclockwise. Therefore, only the outside lines (the heavy lines) remain. If we index the (small) cycles making up the complete cycle, by an index, k, where k ranges over all the cells inside the heavy line, this gives,

$$\sum_{all\,cycle} \frac{q_k}{T_k} = \sum_{outside\,lines,k} \frac{q_k}{T_k} \leq 0 \tag{11}$$

In the limit where we approximate our arbitrary closed path by making the isotherms and adiabatic lines closer and closer together to create an infinite number of cycles, the summation in Equation 11 becomes,

$$\oint_{closed\,path} \frac{dq}{T} \leq 0 \tag{12}$$

The equal sign holds when the entire path is reversibly,

$$\oint_{closed\,path} \frac{dq_{rev}}{T} \leq 0 \tag{13}$$

This equation says that the integral of reversible heat over temperature around a closed path is zero. Equation 13 is independent of the shape of the path - as long as it is a closed loop - which implies that,

$$\frac{dq_{rev}}{T} \tag{14}$$

is the differential of a state function. We will call this state function entropy and give it the symbol, S. Equation 14 becomes

$$dS = \frac{dq_{rev}}{T}. \tag{15}$$

Combining Equations (12), (13), and (15) we get,

$$\oint_{closed\,path} \frac{dq}{T} \le \oint_{closed\,path} \frac{dq_{rev}}{T} = \oint_{closed\,path} dS. \tag{16}$$

Since the closed path is arbitrary, Equation (16) must be valid for any possible closed path. The only way this can be true is if the equation is true in its differential form. That is,

$$\frac{dq}{T} \le \frac{dq_{rev}}{T} = dS, \tag{17}$$

which can be rewritten as simply

$$\frac{dq}{T} \le dS, \tag{18}$$

or, the best way to remember it,

$$dS \ge \frac{dq}{T}, \tag{19}$$

with the understanding that the equal sign holds when the process is reversible and the greater-than sign holds when the process is irreversible.

Equation (19) is the second law of thermodynamics in equation form. Equation (18) would work just as well, but most of us would prefer (19) because it is in the form of a definition of dS.

SECOND LAW APPLICATIONS - EQUILIBRIUM AND ENTROPY CHANGES

The second law of thermodynamics in equation form is,

$$dS \geq \frac{dq}{T}, \tag{1}$$

where the = sign holds when the process is reversible and the > holds when it is irreversible. Our first application of the second law will be to provide a thermodynamic definition of equilibrium. We indicated very early in the course that at equilibrium none of the variables was changing in time. (Of course, a system in a steady state has to be excluded from this definition.) Another possible definition of equilibrium was that all the variables have the values they would have at time equals infinity. Neither of these "definitions" provides an equation we can use to discuss equilibrium systems carefully.

The second law provides us with a definition of equilibrium that we can use to derive equilibrium properties of thermodynamic systems. Consider a closed isolated system. In a closed isolated system $dq = 0$ and $dV = 0$ which implies that $dU = 0$. (Actually, we should really say that dw for all forms of work are zero, but for the time being we will only consider pV work.)

Under the conditions of constant U and V the second law, Equation 1, becomes

$$dS_{UY} \geq 0. \tag{2}$$

This may look simple, but it is a very profound statement. It says that in a closed isolated system (a system which us not being disturbed from the outside) any spontaneous change must increase the entropy.

In a closed, isolated system the entropy seeks a maximum.

If you plot the entropy of a closed isolated system against some system variable any spontaneous change in the system must take the system to higher entropy. If the system is not in equilibrium then

$$dS_{UY} > 0,$$

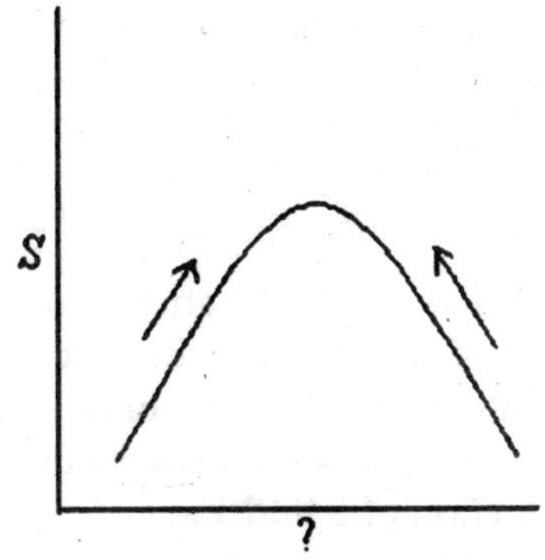

but if the system is at equilibrium any spontaneous change in the system must leave the entropy unchanged,

$dS=0$

Equation 2 is the origin of the somewhat arrogant statement that you may have heard, "The entropy of the universe is increasing." If you regard the universe as a closed isolated system then the statement is probably true, even though it is difficult to take such broad statements concerning the nature of the universe seriously.

COMBINED FIRST AND SECOND LAWS

The first thing we must do is incorporate our new-found equation for the second law into what we already know. Going back to the first law, with pV work only, recall that we can write

$$dU = dq - pdV. \quad (3)$$

If we restrict our attention to reversible processes this becomes,

$$dU = dq_{rev} - pdV, \quad (4)$$

but we know from the second law that

$$dq_{rev} = Tds. \quad (5)$$

Merging Equations 4 and 5 we arrive at what is called the combined first and second laws,

$$dU = TdS - pdV. \quad (6)$$

(Later on, when we want to include work other than pV work, we will add it in to Equation 6,

$$dU = TdS - pdV + dw_{other}, \tag{7}$$

but for now we will just stick with pV work.)

Equation 6 implies that the natural variables of internal energy, U, are S and V. In previous calculations we have regarded U and a function of T and V or of T and p, but nature - in the form of the first and second laws of thermodynamics - gives us U as a function of S and V.

Divide Equation 6 by dT and hold V constant:

$$\left(\frac{\partial U}{\partial T}\right)_V = T\left(\frac{\partial S}{\partial T}\right)_V + 0 = C_V$$

$$\left(\frac{\partial S}{\partial T}\right)_V = \frac{C_V}{T}. \tag{8a, b}$$

Equation 8b provides us with a way to calculate entropy changes for a certain class of processes, namely processes at constant volume. Set up Equation 8b for integration,

$$dS_V = \frac{C_V}{T}dT, \tag{9}$$

and calculate the entropy change for a constant volume process as,

$$\Delta S_V = \int_{T_1}^{T_2} \frac{C_V}{T}dT. \tag{10}$$

Continuing to incorporate the second law into our set of thermodynamic tools, recall that

$$H = U + pV$$
$$dH = dU + pdV + Vdp$$
$$dH = TdS = pdV + pdV + Vpd$$
$$dH = TdS + Vdp. \tag{11a, b, c, d}$$

(Note that equation 11d implies that the natural variables of H are S and p.)

This time let's divide Equation 11d by dT and hold p constant,

$$\left(\frac{\partial H}{\partial T}\right)_p = T\left(\frac{\partial S}{\partial T}\right)_p + 0 = C_p$$

$$\left(\frac{\partial S}{\partial T}\right)_p = \frac{C_p}{T}. \qquad (12a, b)$$

We can use Equation 12b to calculate entropy changes for processes at constant pressure. Set up Equation 12b for integration,

$$dS_p = \frac{C_p}{T} dT, \qquad (13)$$

and integrate,

$$\Delta S_p = \int_{T_1}^{T_2} \frac{C_p}{T} dT. \qquad (14)$$

What about processes at constant T? We can calculate the entropy change for a process at constant temperature by rearranging Equation 5 for integration,

$$dS = \frac{dq_{rev}}{T}. \qquad (15)$$

The finite entropy change is,

$$\Delta S = \int \frac{dq_{rev}}{T}. \qquad (16)$$

If we now restrict consideration to processes at constant T Equation 16 becomes,

$$\Delta S = \int \frac{dq_{T,rev}}{T}$$

$$= \frac{1}{T} \int dq_{T,rev}$$

$$= \frac{q_{T,rev}}{T}. \qquad (17)$$

Example Calculation

Calculate the entropy change in heating 1.00 mol of Al from 300 K to 500 K at constant pressure. The constant pressure heat capacity of Al is given to good approximation by,

$$C_p = 20.67 \frac{J}{Kmol} + 12.38 \times 10^{-3} T \frac{J}{K^2 mol}.$$

The calculation is

$$\Delta S_V = \int_{T_1}^{T_2} \frac{20.67}{T} dT + \int_{T_1}^{T_2} 12.38 \times 10^{-3} dT$$

$$= 20.67 \, In \frac{500K}{300K} + 12.38 \times 10^{-3} (500K - 200K)$$

$$= 13.04 \frac{J}{K}.$$

Another Example - An irreversible Process

Often we must calculate entropy changes for irreversibly processes. We don't know how to calculate entropy changes for irreversible processes, but it doesn't matter. Entropy is a state function so ΔS is independent of path. All we have to do is imagine a reversible path which will effect the same change and calculate the entropy change for the reversibly path.

Suppose we start with a 100. g block of Cu at 500 K and a 100. g block of Cu at 300 K. Bring the two blocks into thermal contact and heat will flow from the hotter block to the cooler block until they reach the same temperature. This is definitely an irreversible process. However, let's imagine that we can reversibly cool the hot Cu block down to the final temperature and reversibly warm the cold Cu block up to the final temperature. We can calculate the entropy change for both of these processes from Equation (14). The total entropy change is just the sum of the two individual entropy changes. Note that we expect the entropy for this process to be positive because the process is spontaneous and the two Cu-block system can be regarded as isolated.

We need the following data:

$$C_{p,m,Cu} = 22.64\frac{J}{K}$$

$$FW_{Cu} = 63.5\frac{g}{mol}$$

First we must find the final temperature. We do this by recognizing that the heat lost by one Cu block is gained by the other Cu block. This is called "heat balance."

$$\frac{100g}{63.5\frac{g}{mol}}22.64\left(500 - T_{fin}\right) = \frac{100g}{63.5\frac{g}{mol}}22.64\left(T_{fin} - 300\right).$$

This is easily solved to obtain T_{fin} = 400 K. Then

$$\Delta S_1 = \frac{100g}{63.5\frac{g}{mol}}22.64\, In\frac{400K}{500K} = -7.95\frac{J}{K}$$

$$\Delta S_2 = \frac{100g}{63.5\frac{g}{mol}}22.64\, In\frac{400K}{500K} = +10.25\frac{J}{K}$$

$$\Delta S = \Delta S_1 + \Delta S_2 = 2.30\frac{J}{K}.$$

ENTROPY OF MIXING (IDEAL GASES)

Visualize that we have a container divided into two compartments. In one compartment we have n_1 moles of an ideal gas, gas 1, at pressure, p and temperature, T. In the other compartment we have n_2 moles of another ideal gas, gas 2, at the same p and T.

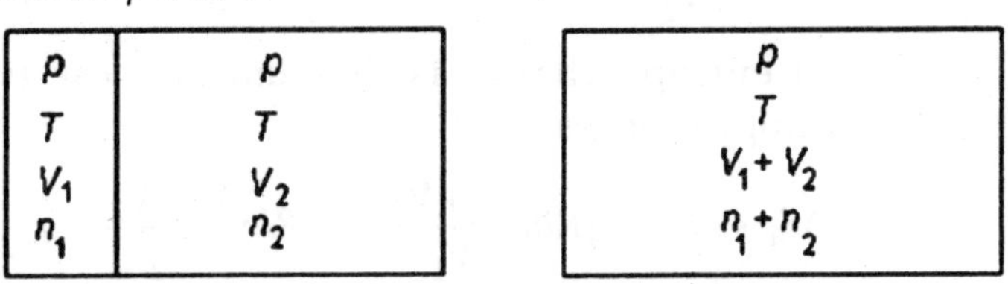

If we remove the partition the gases will begin to diffuse into each other and the system will eventually reach the state where both gases are uniformly distributed throughout the container. This is clearly an irreversibly process so that we would expect that the entropy would increase.

To calculate the entropy change we must find a reversible path to carry out the process, even if the path is fictitious. Imagine that we can devise a process that will expand one gas reversibly and isothermally, but leave the other gas undisturbed. We know how to calculate the change in entropy for the reversible isothermal expansion of an ideal gas.

Recall that $dU = 0$ for the isothermal expansion of an ideal gas. Then,

$$dU = TdS - pdV = 0 \quad (1)$$

So,

$$TdS = pdV$$

$$dS = \frac{p}{T}dV$$

$$= \frac{nR}{V}dV, \quad (2a, b, c)$$

for an ideal gas. So, for gas number 1 in our fictitious isothermal expansion we have,

$$\Delta S_1 = n_1 R\, In \frac{V_1 + V_2}{V_1}, \quad (3a)$$

and for gas number 2,

$$\Delta S_2 = n_2 R\, In \frac{V_1 + V_2}{V_2}. \quad (3b)$$

The total entropy change is the sum of these two individual entropy changes,

$$\Delta S_{mix} = \Delta S_1 + \Delta S_2 = n_1 R In \frac{V_1 + V_2}{V_1} + n_2\, R In \frac{V_1 + V_2}{V_2}. \quad (4)$$

Equation 4 could be used for calculations, but it is not in the form that we will see in other contexts. To obtain the usual form factor the R out of Equation 4 and invert the argument of the logarithms,

$$\Delta S_{mix} = R\left(n_1 In \frac{V_1}{V_1 + V_2} + n_2\, In \frac{V_2}{V_1 + V_2} \right). \tag{5}$$

Since we are doing a calculation for ideal gases, notice that the argument of the first logarithm in Equation 5 can be written,

$$\frac{V_1}{V_1 + V_2} = \frac{\dfrac{n_1 RT}{p}}{\dfrac{n_1 RT}{p} + \dfrac{n_2 RT}{p}} = \frac{n_1}{n_1 + n_2} = X_1, \tag{6}$$

where X_1 is the mole fraction of component 1. There is an equivalent expression for the argument of the second logarithm. The entropy of mixing becomes,

$$\Delta S_{mix} = -R\left(n_1 In\, X_1 + n_2 In\, X_2\right). \tag{7}$$

Equation 7 is also suitable for calculations, but it is not yet in the standard form. To obtain the standard form write the total number of moles $n_1 + n_2$ as n and multiply and divide equation 7 by n. The result is

$$\Delta S_{mix} = -nR\left(X_1 In\, X_1 + X_2 In\, X_2\right). \tag{8}$$

Equation 8 is the same form that we will find when we derive an expression for the entropy of mixing ideal solutions and it is the same form that Shannon found for the "entropy of a message" in his famous series of papers on information theory.

If the two gases were not at the same initial pressure you would have to introduce some extra steps. Expand or compress one of the gases to bring it to the pressure of the other gas, mix the gases, and then compress or expand the mixture to bring it to the correct final volume and pressure.

If the two gases are not at the same temperature and pressure it is more complicated. You must find the final temperature using heat balance, reversibly cool and heat the two gases respectively to the same temperature, expand or contract on of the gases, mix them, and then expand or contract the mixture to the appropriate volume.

Equation 8 can easily be extended to more than two gases,

$$\Delta S_{mix} = -nR\left(X_1 In\, X_1 + X_2 In\, X_2 + X_3 In\, X_3 + ...\right). \quad (9)$$

Example - The molar entropy of dry air

The composition of dry air is approximately 78 per cent N_2, 21 per cent O_2, and 1 per cent Ar by volume (which is the same as mole per cent). What is the molar entropy of mixing of air?

$$\Delta S_{mix,air} = 1 \times mol \times 8.314 \frac{J}{K\,mol}$$

$$\left(0.78\, In\, 0.78 + 0.21\, In\, 0.21 + 0.01\, In\, 0.01\right)$$

$$= 4.7 \frac{J}{K}.$$

ENTROPY MEASURES DISORDER

If we look at the processes we have seen which have positive entropy changes we can see that in each case an increase in entropy is associated with an increase in disorder.

An isothermal expansion gives the molecules more room to move around in, the molecules are less localized.

Increasing the temperature increases the average speeds of the molecules. The molecules are said to be more disordered in "velocity space" (or momentum space).

Mixing gases (or liquids) intersperses the molecules among each other increasing the disorder.

Phase changes, such as going from a solid to a liquid or a gas, or from a liquid to a gas, increase the entropy because gases are more disordered than solids or liquids and liquids are more disordered than solids.

For example, the entropy of fusion of 1.00 mol of ice at 273.15 (heat of fusion is 6.008 kJ/mol) is

$$\Delta S_{fus} = \frac{6008J}{273.15K} = 22.0\frac{J}{K}.$$

Vapourization of liquids has a large positive entropy of vapourization because gases are greatly disordered compared to liquids. A typical value is obtained from the vapourization of benzene at its boiling point. The heat of vapourization is 30.8 kJ at the boiling point, 353.1K. The entropy of vapourization is

$$\Delta S_{Vap} = \frac{30800J}{353.1K} = 87.2\frac{J}{K}.$$

It is interesting that the entropy of vapourization of many substances at their boiling points is close to about 86 J/K. (Water and helium are exceptions.) This phenomenon is called "Trouton's rule." It is easily understood on the basis of entropy being a measure of disorder. The vapourization process essentially "creates" a mole of disordered molecules (the gas) from a mole of highly ordered molecules (the solid or liquid). The gases are all at one atmosphere pressure because we are at their normal boiling point.

Chapter 6

Ideal Eutectic Phase Diagrams Chemistry

The ideal lead-antimony eutectic phase diagram is shown in the figure, and was calculated assuming that the solids are completely immiscible and that the solution was ideal. The liquid-solid coexistence lines are computed from the equilibrium conditions:

$$\mu_i^{(0)}(solid,T) = \mu_i^{(0)}(liquid,T) + RT\,In(x_i), i = A, B.$$

or using the fact that $\Delta G_{fus}^{-(0)}(i) = \mu_i^{(0)}(liquid,T) - \mu_i^{(0)}$

$(solid,T) = 0$ at the normal melting point, $T = T_{MP,}^{(i)}$ of material

i, we can rewrite Eq. (1) as

$$In(x_i) = \int_{T_{MP}^{i}}^{T} \frac{\Delta\overline{H}_{fus}^{(i)}}{RT^2} \approx \frac{\Delta\overline{H}_{fus}^{(i)}}{R}\left[\frac{1}{T_{MP}^{(i)}} - \frac{1}{T}\right]$$

where the last approximation follows by assuming the the heat of

fusion is independent of temperature.

The point where the two curves cross is called the Eutectic point. The question is what happens below the eutectic temperature? From the point of view of the phase rule, or the equilibrium conditions given in Eqs. (1) or (2), we could still have L+A, L+B, or A+B (i.e., a mixture of the solid phases). Indeed, it is the last possibility which happens;

In order to answer this question, we must go back a few steps and remember that under constant T,P,N conditions, the equilibrium state is the one with the *lowest* Gibbs free energy; specifically if there are multiple local minima in G, it is the one that is lowest that should be observed at equilibrium. This is what happens below the eutectic temperature. To see this, consider the total free energy for the system (again for ideal solutions and immiscible solids):

$$G = n_A^{(s)} \mu_A^{(0)}(solid, T) + n_B^{(s)} \mu_B^{(0)}(solid, T)$$

$$+ n_A^{(l)} \left[\mu_A^{(0)}(liquid, T) + RT\, In(x_A) \right] + n_B^l$$

$$\left[\mu_B^{(0)}(liquid, T) + RT\, In(x_B) \right]$$

$$N_A \mu_A^{(0)}(solid, T) + N_B \mu_B^{(0)}(solid, T)$$

$$n_A^{(l)} \left[\Delta G_{fus}^{-(0)}(A, T) + RT\, In(x_A) \right] + n_B^{(l)}$$

$$\left[\Delta G_{fus}^{-(0)}(B, T) + RT\, In(x_B) \right],$$

where n_i^s, n_i^l and N_i are the number of moles of in the solid, liquid, and total, respectively. Remembering that the total number of moles of each material remains constant, the equilibrium states must be local minima of G, and these are found by taking derivatives of Eq. (3b) with respect to $n_A^{(l)}$ or $n_b^{(l)}$ and setting the result to zero. It is straightforward to show that this results in

Eq. (1).

As was discussed above, when deciding between different possibilities, it is necessary to go back to Eq. (3) and find the local minimum with the lowest free energy. For our example, consider the following two possibilities L+A or A+B. The latter obeys the liquid-solid equilibrium condition for A, cf. Eq. (1) and hence, from Eq. (3b) we see that the states have free energies:

$$G_{A+B} = N_A \mu_A^{(0)}(solid,T) + N_B \mu_B^{(0)}(solid,T) \quad (4a)$$

and

$$G_{L+A} = N_A \mu_A^{(0)}(solid,T) + N_B \mu_B^{(0)}(solid,T) + N_B\left[\Delta \bar{G}_{fus}^{(0)}(B,T) + RT\,In(x_B)\right], \quad (4b)$$

where we have set $n_A^{(l)} = n_B^{(l)} = 0$ for the A+B state and $n_B^{(l)} = N_B^{(l)}$ for the L+A state. Thus, L+A state will be the equilibrium one as long as

$$\frac{\Delta \bar{G}_{fus}^{(0)}(B,T)}{RT} + In(x_B) < 0 \quad (5)$$

or, approximately, as long as

$$In(x_B) < \frac{\Delta \overline{H}_{fus}}{R}\left(\frac{1}{T_{MP}^{(B)}} - \frac{1}{T}\right)$$

which, cf. Eq. (2), breaks down below the L+B coexistence line. Thus, since all points on the L+A coexistence line are below the L+B coexistence line below the eutectic temperature, there the L+A state has a higher free energy than A+B state. Clearly, there is nothing special about the notation for A, and hence, exactly the same conclusion can be drawn for B. Thus we see that the solid mixture is the minimum free energy solution below the eutectic temperature. Finally, while we have used ideal solutions in this example, you can show that the same argument still holds for a nonideal solution

Chapter 7

Capillary Rise and Depression

Consider a two phase system with a flat interface as shown in figure.

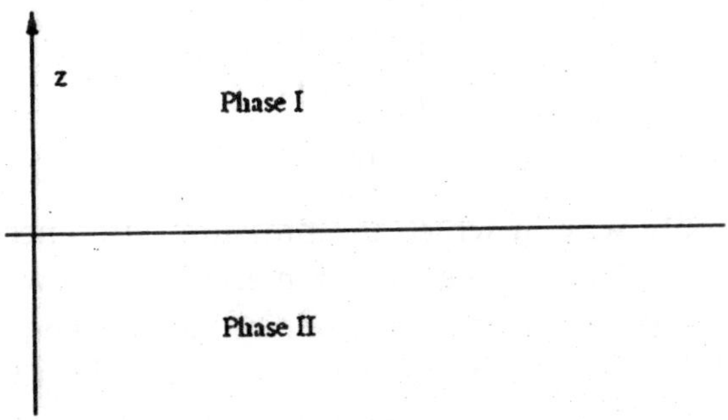

Fig.

Since the two phases are in equilibrium in the presence of a gravity field, the pressure varies with height, z; namely,

$$p_1(z) = p_1(0) - \rho_1 g z \quad 1$$

in phase 1, and

$$p_2(z) = p_2(0) - \rho_2 g z \quad 2$$

in phase ρ1(2), where is the density in phase 1(2). Of course, at the interface (here assumed flat) the pressures must be equal and

$$p_1(0) = p_2(0) = p_0. \quad 3$$

Now put a capillary tube into the system as shown in figure.

The surface tension, γ, between phase 1 and phase 2 wants the pressures between the two phases to be unequal at the curved interface (why?). Specifically, for the convex surface shown in figure we know that

$$p_{2(-h)} - p_1(-h) = \frac{2\gamma}{R}. \tag{4}$$

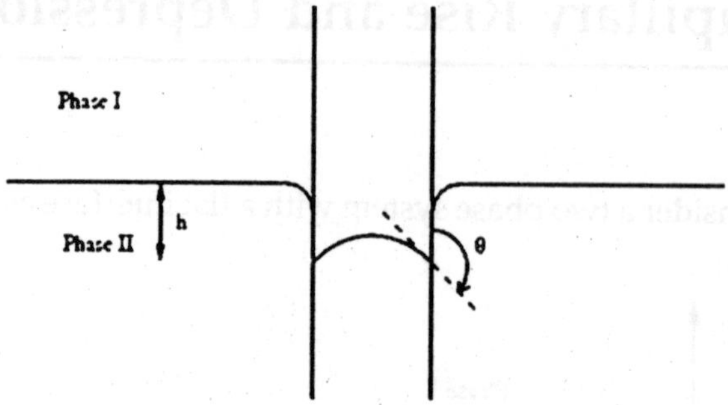

Fig.

The only way in which the pressure difference can be achieved is by moving the curved interface to a new position in the fluid. (Changing the pressure in the gas or liquid from that of the surrounding material at the same height would cause material to flow from the high to low pressure regions until the pressures were equal at all points at the same height). Thus, by using our expressions for the pressure dependence in the phases at any given heights [Eqs. (1), (2), and (3)] in Eq. (4) we find that

$$p_0 + \rho_2 gh - (p_0 + \rho_1 gh) = \frac{2\gamma}{R} \tag{5}$$

or

$$h = \frac{2\lambda}{Rg(\rho_2 - \rho_1)} \approx \frac{2\gamma}{Rg\rho_2} \text{ for } \rho_2 >> \rho_1, \tag{6}$$

where the approximate expression is useful for liquid-vapour interfaces. (If the interface is concave then the pressures in Eq. (4) must be switched and the sign of changes.)

Finally, if we write R, the radius of curvature of the interface, in terms of d, the radius of the tube and the contact angle, it follows that

$$h = -\frac{2\gamma\cos(\theta)}{dg(\rho_2 - \rho_1)}$$

For a 45° contact angle in a capillary tube with a $1mm$ radius containing water ($\gamma = 72.75\,dynes/cm$ the capillary *rise* will be 1.05cm.

SOME TOOLS OF THERMODYNAMICS

SOME MISCELLANEOUS RELATIONSHIPS

Recall that the combined first and second laws give the relationship

$$dU=TdS-pdV. \tag{1}$$

This implies that U is a function of S and V. Sometimes we call S and V the "natural variables" of U. Regarding $U = U(S,V)$ we can write

$$dU = \left(\frac{\partial U}{\partial S}\right)_V dS + \left(\frac{\partial U}{\partial V}\right)_S dV. \tag{2}$$

Comparing Equations 1 and 2 it is clear that

$$\left(\frac{\partial U}{\partial S}\right)_V = T, \tag{3a}$$

and

$$\left(\frac{\partial U}{\partial V}\right)_S = -p. \tag{3b}$$

These two equations can be regarded as thermodynamic definitions of T and p.

Likewise, from the definition of enthalpy we wrote before that

$$dH=TdS+Vdp. \tag{4}$$

Equation 4 implies that enthalpy is a natural function of S and p. Regarding $H = H(S,p)$ we can write

$$dH = \left(\frac{\partial H}{\partial S}\right)_p dS + \left(\frac{\partial H}{\partial p}\right)_S dp \tag{5}$$

From Equations 4 and 5 it is clear that

$$\left(\frac{\partial H}{\partial S}\right)_p = T, \tag{6a}$$

and

$$\left(\frac{\partial H}{\partial p}\right)_S = V. \tag{6b}$$

Equations 6a and 6b give us another thermodynamic definition of T and a thermodynamic definition of V (which is curious since we have always regarded V as a purely mechanical variable).

HELMHOLTZ AND GIBBS FREE ENERGY

When we made the transformation from U to H by the Legendre transformation,

$$H=U+pV,. \tag{7}$$

we remarked that V was not the most convenient independent variable. In the laboratory it is usually much easier to control pressure than it is to control V. Since both U and H are natural functions of entropy, it is fair to ask how convenient it is to have S as a variable. The answer is that it is not at all convenient to control entropy or to have entropy as an independent variable. We do not have a meter that reads entropy and we do not know how to hold entropy constant as we change some other variable. (Recall that we can control temperature, pressure, and volume.) So we make some more Legendre transformations.

(We will not give an extensive discussion of Legendre transformations here, but we should point out that they are not arbitrary. You can't just pick any two variables you wish and put them together to make a Legendre transformation.

We could make the Legendre transformation from U *to* H by adding the pV term to U only because V is related to p and U through Equation 3b. Using this as a guide it would seem reasonable to use Equation 3a to change the variable S to T in the function U, and use Equation 6a to change the variable S to T in the function H. Let's try it.)

Define the Helmholtz free energy, A, as

$$A \equiv U - TS. \tag{8}$$

Then,

$$dA = dU - TdS - SdT$$
$$= TdS - pdV - TdS - SdT$$
$$= -SdT - pdV. \tag{9a, b, c}$$

Equation 9c tells us that the Helmholtz free energy is a natural function of T and V. That is, $A = A(T,V)$. T and V are much more convenient variables than S and V. Regarding $A = A(T,V)$ we see that

$$dA = \left(\frac{\partial A}{\partial T}\right)_V dT + \left(\frac{\partial A}{\partial V}\right)_r dV. \tag{10}$$

Now compare Equation 10 with Equation 9c to see that,

$$\left(\frac{\partial A}{\partial T}\right)_V = -S, \tag{11a}$$

and

$$\left(\frac{\partial A}{\partial V}\right)_r = -p. \tag{11b}$$

Equation 11a gives us a thermodynamic definition of entropy and 11b gives another thermodynamic definition of pressure.

We now have a function of T and V, but we didn't much like V as an independent variable before so why should we like it any better now? Let's use the relationship 6a to define the Gibbs free energy, G, as,

$$G \equiv H - TS. \tag{12}$$

Actually, we can use any one of the three equivalent definitions,

$$G = H - TS$$
$$G = A + pV$$
$$G = U + pV - TS. \qquad (13a, b, c)$$

Any one of Equations 13a, b, or c will give us the correct natural variables of G. Use Equation 12.

$$dG = dH - TdS - SdT$$
$$= TdS + Vdp - TdS - SdT$$
$$= -SdT + Vdp. \qquad (14a, c, c)$$

From Equation 14c we see that the natural variables of G are temperature and pressure. Write

$$dG = \left(\frac{\partial G}{\partial T}\right)_P dT + \left(\frac{\partial G}{\partial p}\right)_r dp. \qquad (15)$$

Comparing Equation 15 with Equation 14c we find that.

$$\left(\frac{\partial G}{\partial T}\right)_P = -S, \qquad (16a)$$

and

$$\left(\frac{\partial G}{\partial p}\right)_r = V. \qquad (16b)$$

Equation 16a gives us another thermodynamic definition of entropy and 16b another definition of volume.

Meaning of *A* and *G*

What do *A* and *G* mean and what are they good for? We said after the introduction of the first law (which introduced the internal energy, *U*) that we would be introducing three more functions that have units of energy. We now know that these functions are *H*, *A*, and *G*. At the time we said that only *U* has a simple physical meaning - the sum of all the kinetic and potential energies of all the particles. There is no simple physical explanation for enthalpy and the two free energies. The best we can do is tell how they are used.

1. Most simple-minded.

Set

$$dU = TdS - pdV + dw_{other}- \tag{17}$$

Then, using the definition of Helmholtz free energy as we have done above we find that,

$$dA = -SdT - pdV + dw_{other}- \tag{18}$$

For any process at constant temperature we have,

$$dA_T = -pdV + dw_{other} = dw_{all,reversible}\cdot \tag{19}$$

That is, for a constant temperature process the Helmholtz free energy gives all the reversible work. For this reason the Helmholtz free energy is sometimes called the "work function." When a physicist says "free energy" without indicating Helmholtz or Gibbs, he usually means Helmholtz free energy.

Similarly, we can write,

$$dG = -SdT + Vdp \_ dw_{other}\cdot \tag{20}$$

For a process at constant temperature *and* pressure we get,

$$dG_{T,p} = dw_{other} = dw_{all,non-pV}\cdot \tag{21}$$

That is, for a process at constant temperature and pressure the change in Gibbs free energy gives all the reversible work except the pV work. This work might include electrical work, work creating surface area, and so on. Chemists do most of their reactions on the bench top at constant pressure. When a chemist says "free energy" she almost always means Gibbs free energy unless she specifically states otherwise.

2. More useful - two new criteria for equilibrium

Recall that the second law of thermodynamics,

$$dS \geq \frac{dq}{T}, \tag{22}$$

gives us the fundamental criterion for equilibrium. That is, in a closed isolated system entropy seeks a maximum,

$$dS_{U,V} \geq 0 \tag{23}$$

Although this is the fundamental definition of equilibrium it is not the most useful definition because we do not often work with closed isolated systems. More often we work with systems at constant temperature and either constant volume or constant pressure. We can use the second law, Equation 22, and our new functions *A* and *G* to find criteria for equilibrium under these conditions.

Rewrite the second law, Equation 22 as follows:

$$TdS \geq dq$$

$$0 \geq dq - TdS, \tag{24}$$

or

$$dq - TdS \leq 0. \tag{25}$$

Now, going back to the original form of the first law with only *pV* work,

$$dU = dq - pdV, \tag{26}$$

and making the transformation to Helmholtz free energy we get,

$$\begin{aligned} dA &= dU - TdS - SdT \\ &= dq - pdV - TdS - SdT \\ &= (dq - TdS) - pdV - SdT. \end{aligned} \tag{27a, b, c}$$

For a process at constant temperature and volume we have,

$$dA_{TV} = dq - TdS \leq 0 \tag{28}$$

We conclude that for a process at constant temperature and volume the Helmholtz free energy seeks a minimum. Any spontaneous process in a system at constant *T* and *V* must decrease the Helmholtz free energy (if the system is away from equilibrium) or leave the Helmholtz free energy unchanged (if the system is at equilibrium).

By the same token, we can use the Gibbs free energy to discuss processes at constant temperature and pressure,

$$\begin{aligned} dG &= dH - TdS - SdT \\ &= dq + Vdp - TdS - SdT \\ &= (dq - TdS) + Vdp - SdT, \end{aligned} \qquad \text{(29a, b, c)}$$

from which we conclude that for a process at constant T and p,

$$dG_{T,p} = dq - TdS \leq 0. \qquad (30)$$

That is, at constant T and p the Gibbs free energy seeks a minimum. Any spontaneous process in a system at constant T and p must decrease the Gibbs free energy (if the system is away from equilibrium) or leave the Gibbs free energy unchanged (if the system is at equilibrium).

MAXWELL'S EQUATIONS

We now have the tools to derive some very useful relationships between thermodynamics variables. Maxwell's equations are based on the same principle as was Euler's test for exact differentials, namely that mixed second derivatives of "nice" functions must be equal. Applying this principle to our two new free energy functions we find,

$$\frac{\partial^2 A}{\partial V \partial T} = \frac{\partial^2 A}{\partial T \partial V}, \qquad (31)$$

but we already know the first derivatives of A from Equations 11a and 11b. So,

$$\begin{aligned} &\frac{\partial^2 A}{\partial V \partial T} = \frac{\partial}{\partial V}\left(\frac{\partial A}{\partial T}\right)_V = \frac{\partial^2 A}{\partial T \partial V} = \frac{\partial}{\partial T}\left(\frac{\partial A}{\partial V}\right)_T, \\ &\frac{\partial}{\partial V}\left(\frac{\partial A}{\partial T}\right)_V = \frac{\partial}{\partial V}(-S) = \frac{\partial}{\partial T}\left(\frac{\partial A}{\partial V}\right)_T = \frac{\partial}{\partial T}(-p), \\ &\left(\frac{\partial S}{\partial V}\right)_T = \left(\frac{\partial p}{\partial T}\right)_V. \end{aligned} \qquad \text{(32a, b, c)}$$

We obtain another similar equation from the Gibbs free energy,

$$\frac{\partial^2 G}{\partial p \partial T} = \frac{\partial^2 G}{\partial T \partial p}, \tag{33}$$

which becomes, using Equations 16a and 16b,

$$\frac{\partial^2 G}{\partial p \partial T} = \frac{\partial}{\partial p}\left(\frac{\partial G}{\partial T}\right)_p = \frac{\partial^2 G}{\partial T \partial p} = \frac{\partial}{\partial T}\left(\frac{\partial G}{\partial p}\right)_T,$$

$$\frac{\partial}{\partial p}\left(\frac{\partial G}{\partial T}\right)_p = \frac{\partial}{\partial p}(-S) = \frac{\partial}{\partial p}\left(\frac{\partial G}{\partial p}\right)_T = \frac{\partial}{\partial T}(+V),$$

$$-\left(\frac{\partial G}{\partial p}\right)_T = \left(\frac{\partial V}{\partial T}\right)_p. \tag{34a, b, c}$$

There are two more Maxwell's equations from dU and dH, but these are not as useful as the ones just derived. We will leave it to the reader to find the Maxwell's equations from dU and dH.

FIRST APPLICATION OF A MAXWELL'S EQUATION

As our first application of a Maxwell's equation we will derive the so-called thermodynamic equation of state which we stated without proof earlier. Write the combined first and second laws,

$$dU = TdS - pdV. \tag{1}$$

Divide Equation 1 by dV and hold T constant to get,

$$-\left(\frac{\partial U}{\partial V}\right)_T = T\left(\frac{\partial S}{\partial V}\right) - p. \tag{35}$$

Using the Maxwell's Equation, Equation 32c, to substitute for the entropy derivative we obtain,

$$\left(\frac{\partial U}{\partial V}\right)_T = T\left(\frac{\partial p}{\partial T}\right)_V - p. \tag{36}$$

Equation 36 is the equation that was written down without proof at the time we were discussing the Joule expansion.

The other version of thermodynamic equation of state, based on H instead of U, will be left as an exercise for the reader.

We now have four interesting and useful derivatives of entropy,

$$\left(\frac{\partial S}{\partial V}\right)_T = \frac{C_V}{T} \left(\frac{\partial S}{\partial T}\right)_p = \frac{C_p}{T}$$

$$\left(\frac{\partial S}{\partial V}\right)_T = \left(\frac{\partial p}{\partial T}\right)_V \left(\frac{\partial S}{\partial p}\right)_T = -\left(\frac{\partial V}{\partial T}\right)_p .$$

There are two other derivatives of entropy which might prove useful,

$$\left(\frac{\partial S}{\partial p}\right)_V = \left(\frac{\partial S}{\partial T}\right)_V \left(\frac{\partial T}{\partial p}\right)_V$$

$$= \left(\frac{C_V}{T}\right)\left(\frac{\partial T}{\partial p}\right)_V . \qquad (37)$$

The other derivative, $\left(\frac{\partial S}{\partial V}\right)_p$, will be left to the reader.

ADIABATIC COMPRESSIBILITY

The speed of sound in a gas depends on the "springiness" of the gas. That is, it depends on how the volume of the gas responds to changes in pressure. We have already seen one measure of this response, called the isothermal compressibility,

$$K_T = -\frac{1}{V}\left(\frac{\partial V}{\partial p}\right)_T . \qquad (1)$$

Equation one gives a parameter that determines how the gas responds to changes in pressure if the temperature remains constant.

Sir Isaac Newton assumed that the speed of sound was an isothermal process and used the parameter defined by Equation 1 to calculate the speed of sound in a gas. His answer did not agree with experiment.

It turns out that sound transmission in a gas is an adiabatic process rather than an isothermal process. The sound wave causes oscillations in pressure but the oscillations are fast enough that heat can not move from compressed regions to rarified regions in order to keep the temperature constant. Before the heat can be conducted away from the compressed regions the compression has moved on so that sound propagation is adiabatic.

We define the adiabatic compressibility as,

$$K_S = -\frac{1}{V}\left(\frac{\partial V}{\partial p}\right)_S . \tag{2}$$

We can calculate the adiabatic compressibility in terms of quantities that we already know (using the Euler cyclic rule twice, once the normal way and once in reverse).

$$\begin{aligned} k_S &= -\frac{1}{V}\left(\frac{\partial V}{\partial p}\right)_S \\ &= -\frac{1}{V}\left(-\frac{\left(\frac{\partial S}{\partial p}\right)_V}{\left(\frac{\partial S}{\partial V}\right)_p}\right) \\ &= -\frac{1}{V}\left(-\frac{\left(\frac{\partial S}{\partial T}\right)_V\left(\frac{\partial T}{\partial p}\right)_V}{\left(\frac{\partial S}{\partial T}\right)_p\left(\frac{\partial T}{\partial V}\right)_V}\right), \end{aligned} \tag{3a, b, c}$$

where we have used Euler's cyclic relation to go from Equation 3a to 3b and the chain rule in the denominator and numerator of 3b to go to 3c.

We already know that

$$\left(\frac{\partial S}{\partial T}\right)_V = \left(\frac{C_V}{T}\right) \tag{4a}$$

and

$$\left(\frac{\partial S}{\partial T}\right)_p = \left(\frac{C_p}{T}\right), \tag{4b}$$

so

$$k_S = -\frac{1}{V}\left(-\frac{\left(\frac{C_V}{T}\right)\left(\frac{\partial T}{\partial p}\right)_V}{\left(\frac{C_p}{T}\right)\left(\frac{\partial T}{\partial V}\right)_p}\right)$$

$$= \frac{C_V}{C_p}\frac{1}{V}\left(\frac{\left(\frac{\partial T}{\partial p}\right)_V}{\left(\frac{\partial T}{\partial V}\right)_p}\right). \tag{5a, b}$$

In Equation 5b we can use Euler's cyclic rule in reverse to write

$$k_S = \frac{C_V}{C_p}\frac{1}{V}\left(\frac{\partial V}{\partial p}\right)_T$$

$$k_S = \frac{C_V}{C_p}k_S. \tag{6a,b}$$

Since $C_p > C_V$ The isothermal compressibility is always greater than the adiabatic compressibility.

For a monatomic ideal gas, where $C_p = 5nR/2$ and $C_V = 3nR/2$ we see that

$$kS = \frac{3}{5}k_T. \tag{7}$$

ADIABATIC GAS EXPANSION REVISITED

Early in the course we derived the equation for the adiabatic expansion of an ideal gas,

$$pV^{\gamma} = constant,$$

$$p_1V_1^{\gamma} = p_2V_2^{\gamma}, \tag{8a, b}$$

where,

$$\gamma = \frac{C_p}{C_V}. \tag{9}$$

You may recall that the derivation was sort of "round-about." Here we would like to use some of our new thermodynamics tools to provide a much more direct derivation.

The goal is to discover how the pressure changes with volume if entropy is held constant. That is, we would like to find the derivative, $\left(\frac{\partial p}{\partial S}\right)_S$.

Then we can integrate this derivative to find an expression for p as a function of V. First, let's find the derivative. Notice that this derivative is just the reciprocal of the derivative in Equation 2 so most of the work has already been done. Using the same procedures we used above we fin that,

$$\left(\frac{\partial p}{\partial V}\right)_S = -\frac{\left(\frac{\partial S}{\partial V}\right)_p}{\left(\frac{\partial S}{\partial p}\right)_V}$$

$$= -\frac{\left(\frac{\partial S}{\partial T}\right)_p \left(\frac{\partial T}{\partial V}\right)_p}{\left(\frac{\partial S}{\partial T}\right)_V \left(\frac{\partial T}{\partial p}\right)_V}$$

$$= -\frac{\frac{C_p}{T}\left(\frac{\partial T}{\partial V}\right)_p}{\frac{C_V}{T}\left(\frac{\partial T}{\partial p}\right)_V} = \frac{C_p}{C_V}\left(\frac{\left(\frac{\partial T}{\partial V}\right)_p}{\left(\frac{\partial T}{\partial p}\right)_V}\right)$$

$$\left(\frac{\partial p}{\partial V}\right)_S = \gamma\left(\frac{\partial p}{\partial V}\right)_T. \qquad \text{(10a, b, c, d, e, f)}$$

Equation 10f is a general thermodynamic relationship. It contains no approximations. To proceed further we must decide what material we want to consider. The equation we were trying to derive was based on the ideal gas for which

$$\left(\frac{\partial p}{\partial V}\right)_T = -\frac{nRT}{V^2}$$

$$= -\frac{p}{V}. \qquad \text{(11a, b)}$$

So,

$$\left(\frac{\partial p}{\partial V}\right)_S = -\gamma\frac{p}{V}. \qquad (12)$$

Set up Equation 12 for integration, (don't forget to put all the p stuff on one side and all the V stuff on the other)

$$\frac{dp}{p} = -\gamma\frac{dV}{V}, \qquad (13)$$

and integrate between p_1, V_1 and p_2, V_2,

$$\int_{p_1}^{p_2}\frac{dp}{p} = -\gamma\int_{V_1}^{V_2}\frac{dV}{V}, \qquad (14)$$

where we have made the approximations that γ is independent of volume and factored it out of the integral. Equation 14 integrates to

$$In\frac{p_2}{p_1} = -\gamma In\frac{V_2}{V_1}$$

$$= +\gamma In\frac{V_1}{V_2}$$

$$In\frac{V_1^{\gamma}}{V_2^{\gamma}}, \qquad (15a, b, c)$$

from which we conclude that

$$\frac{p_2}{p_1}=\frac{V_1^{\gamma}}{V_2^{\gamma}}, \qquad (16)$$

or

$$p_1V_1^{\gamma}=p_2V_2^{\gamma}. \qquad (16b)$$

Just for the record, we add that the correct formula for the speed of sound is

$$v_{sound}=\frac{1}{\sqrt{\rho k_S}}, \qquad (17)$$

where ñ is the density of the gas (in kg/m^3).

GIBBS FREE ENERGY AND CHEMICAL REACTIONS

We have seen that the Gibbs free energy, *G*, seeks a minimum at constant *T* and *p*. The question now is does this have anything to do with chemical reactions. We define the change in Gibbs free energy for a chemical reaction as,

$$\Delta_r G^{\circ}=G_{products}-G_{reac\tan ts}. \qquad (1)$$

It is understood in Equation 1 that both the reactants and the products are pure, isolated, in their standard states, and at the same temperature and pressure.

When we first introduced heats of reaction we said that it is not feasible to tabulate the ΔH° for every possible reaction. Instead, we tabulate heats of formation for compounds and use Hess' law to find the heat of a reaction involving those compounds.

In the same manner we define the Gibbs free energy of formation for a compound as ΔG° for the reaction:

Pure isolated elements in their standard states → one mole of pure compound in its standard state.

For example, $\Delta_f G^\circ$ for $CO_2(g)$ at 25° and 1 atm is defined to be the ΔG for the reaction:

$$C(s, graphite) + 2\ O_2(g) \rightarrow CO_2(g).$$

Then, using Hess' law, the $\Delta_r G^\circ$ for some arbitrary reaction, say

$$a\,A + b\,B \rightarrow c\,C + d\,D, \quad (2)$$

is

$$\Delta_r G^\circ = c\,\Delta_f G_C{}^\circ + d\,\Delta_f G_D{}^\circ - a\,\Delta_f G_A{}^\circ - b\,\Delta_f G_B{}^\circ. \quad (3)$$

Rarely one might want to run a reaction at constant volume and, therefore, would need the change in the Helmholtz free energy. We can obtain $\Delta_r A^\circ$ using the same approximations we used to obtain ΔU from ΔH. That is, from

$$G = A + pV$$

we see that,

$$\Delta A = \Delta G - \Delta(pV).$$

Using, as before, the approximation that the change in the pV product for liquids and solids is small and the approximation that the gases are ideal, we obtain the equation,

$$\Delta_r A^\circ \approx \Delta_r G^\circ - RT\Delta n_{gas}. \quad (4)$$

PROCESSES AT CONSTANT TEMPERATURE

Knowing that $G = H - TS$, and that reactions are constant temperature process, we can relate the Gibbs free energy of a reaction to the enthalpy of a reaction. That is,

$$G = H + TS,$$

$$\Delta G = \Delta H + \Delta(TS)$$

$$\Delta G_T = \Delta H_T - T\Delta S_T. \quad (5a, b, c)$$

We have written a subscript, T, in Equation 5c to indicate that the equation is only valid at constant T, but it is unusual to include the T. Usually we see

$$\Delta G = \Delta H - T\Delta S, \quad (6)$$

and it is left to the reader to understand that the equation is only valid at constant T.

Recall that at constant T and p the Gibbs free energy seeks a minimum. That means that we can use $\Delta_r G^{\circ}$ to tell whether or not a reaction will proceed spontaneously as written:

If $\Delta_r G^{\circ} > 0$ then the reaction will not go as written (the reverse reaction will go)

if $\Delta_r G^{\circ} < 0$ then the reaction will go as written.

Sometimes people say that $\Delta_r G^{\circ}$ is a measure of the "driving force" of a chemical reaction. Although the use of the word "force" is probably not appropriate, the statement conveys the correct idea that $\Delta_r G^{\circ}$ tells us whether or not a given reaction will really run spontaneously. (You can probably write and balance a large number of reactions, but just because you can write and balance a reaction is no guarantee that it will run.)

We see from Equation 6 that the "driving force" of a chemical reaction has two components:

ΔH is the drive toward stability. When $\Delta H < 0$ the products are more stable than the reactants (and vice versa).

ΔS is the drive toward disorder. When $\Delta S > 0$ the products are more disordered than the reactants.

(The negative sign in front of the $T\Delta S$ shows that a positive ΔS makes a negative contribution to ΔG which tends to drive the reaction in the forward direction.)

Note that increasing T increases the influence of ΔS on the reaction "driving force."

For a chemical reaction ΔH and ΔS are independent of each other. that is, you can not calculate one from the other. You can have situations where both are positive, both are negative, or one is positive and the other negative. Notice that if ΔH and ΔS have the same sign they are working against each other. You can make the entropy win by increasing the temperature or you can make the enthalpy win by decreasing the temperature.

We will show later that $\Delta_r G^{\,o}$ is related to the equilibrium constant for the reaction (as you might expect).

THE THIRD LAW OF THERMODYNAMICS

If we have sufficient heat capacity data (and the data on phase changes) we could write

$$S(T) = S(T=0) + \int_0^T \frac{C_P}{T} dT. \qquad (1)$$

(If there is a phase change between 0 K and T we would have to add the entropy of the phase change.) If C_p were constant near $T = 0$ we would have,

$$S(T) = S(T=0) + C_p In \frac{T}{0},$$

which is undefined. Fortunately, experimentally $C_p \to 0$ as $T \to 0$. For nonmetals C_p is proportional to T^3 at low temperatures. For metals C_p is proportional to T^3 at low temperatures but shifts over to being proportional to T at extremely low temperatures. (The latter happens when the atomic motion "freezes out" and the heat capacity is due to the motion of the conduction electrons in the metal.)

Equation 1 could be used to calculate absolute entropies for substances if we knew what the entropy is at absolute zero. Experimentally it appears that the entropy at absolute zero is the same for all substances. The third law of thermodynamics codifies this observation and sets

for all elements and compounds in their most stable and perfect crystalline state at absolute zero and one atmosphere pressure. (All except for helium, which is a liquid at the lowest observable temperatures at one atmosphere.)

The advantage of this law is that it allows us to use experimental data to compute the absolute entropy of a substance. For example, suppose we want to calculate the absolute entropy of liquid water at 25° C. We would need to know the C_p of ice from 0 K to 273.15 K and the C_p of liquid water from 273.15 K to 298.15 K. We also need the heat of fusion of water at its normal melting point. With all of this data, which can be obtained partly from theory and partly from experiment, we find

$$S^o_{H_2O}(25°C) = 0 + \int_0^{273.15} \frac{C_P(s)}{T} dT + \frac{\Delta H_{fus}}{273.15}$$

$$+ \int_{273.15}^{298.15} \frac{C_p(l)}{T} dT. \quad (2)$$

Some substances may undergo several phase changes.

ENTROPY CHANGES IN CHEMICAL REACTIONS

We can use the third law entropies to calculate entropy changes for chemical reactions. For a typical reaction,

$$a\,\mathrm{A} + b\,\mathrm{B} \rightarrow c\,\mathrm{C} + d\,\mathrm{D}. \quad (3)$$

the entropy change is

$$\Delta_r S° = cS°_C + dcS°_D - acS°_A - bS°_B. \quad (4)$$

Notice two things:

1. We did not define or use an entropy of formation, $\Delta_f S$ o.

2. $S°_{\text{element}}$ is not zero.

As we have said before, $\Delta_r S°$ and $\Delta_r H°$ are independent of each other. They can not be calculated from each other. They must be calculated from Equation 4 and a comparable equation using heats of formation.

There is another way to calculate $\Delta_r S°$,

$$\Delta_r S° = \frac{\Delta_r H° - \Delta_r G°}{T} \quad (5)$$

As we have seen before, the $\Delta_r G°$ and $\Delta_r H°$ can be calculated from free energies and heats of formation.

GIBBS FREE ENERGY AND TEMPERATURE: THE GIBBS-HELMHOLTZ EQUATION

Frequently we wish to run reactions at temperatures other than 25°C. Since we know that the change in the Gibbs free energy, $\Delta_r G^\circ$, between products and reactants tells us whether or not the reaction will run spontaneously we will need this quantity at the new temperature.

(Reminder: If $\Delta_r G^\circ < 0$ the reaction is spontaneous and if $\Delta_r G^\circ > 0$ the reaction is not spontaneous.)

We also know that there are two components to $\Delta_r G^\circ$. That is, $\Delta_r G^\circ = \Delta_r H^\circ$ " $T\Delta_r S^\circ$, where the $\Delta_r H^\circ$ term is only weakly dependent on temperature, but the $T\Delta_r S^\circ$ term is strongly dependent on temperature due to the presence of the T in the term.

So the question becomes, how do these things balance out? What is the dependence of $\Delta_r G^\circ$ on temperature? The simple answer is obtained from the derivative of G with respect to T.

$$\left(\frac{\partial G}{\partial T}\right)_p = -S. \qquad (1)$$

If we apply Equation 1 to $\Delta_r G^\circ = G^\circ_{\text{products}} - G^\circ_{\text{reactants}}$ we get,

$$\left(\frac{\partial \Delta_r G^\circ}{\partial T}\right)_p = \left(\frac{\partial G^\circ_{product}}{\partial T}\right)_p - \left(\frac{\partial G^\circ_{reac\tan ts}}{\partial T}\right)_p$$

$$= -S^o_{products} - \left(-S^o_{reac\tan ts}\right)$$

$$= -\Delta_r S^\circ. \qquad (2a, b, c)$$

Equation 2c shows that if $\Delta_r S^\circ$ is positive $\Delta_r G^\circ$ decreases with temperature, but if $\Delta_r S^\circ$ is negative $\Delta_r G^\circ$ increases with temperature. This will tell us whether $\Delta_r G^\circ$ increases or decreases with increasing temperature.

However, if we want to know whether or not a reaction is favored by an increase or decrease in temperature we really need to be looking at the equilibrium constant. If the equilibrium constant increases with temperature the reaction becomes more favored, but if the equilibrium constant decreases with temperature the reaction becomes less favored.

We will show later that the equilibrium constant for a chemical reaction depends on $\Delta_r G^{\circ}/T$ and not on $\Delta_r G^{\circ}$ all by itself. (There is a good reason for this which we will discuss below.) The Gibbs-Helmholtz equation addresses the question as to how $\Delta_r G^{\circ}/T$ changes with temperature. The Gibbs-Helmholtz equation can be presented in two different, but equivalent forms. If we are just worrying about G itself the two forms look like,

$$\left(\frac{\partial\left(\frac{G}{T}\right)}{\partial\left(\frac{1}{T}\right)}\right)_p = H, \tag{3a}$$

and

$$\left(\frac{\partial\left(\frac{G}{T}\right)}{\partial T}\right)_p = -\frac{H}{T^2}. \tag{3b}$$

Or, applying the procedure we used in Equations 2a, b, and c, we can write,

$$\left(\frac{\partial\left(\frac{\Delta G}{T}\right)}{\partial\left(\frac{1}{T}\right)}\right)_p = \Delta H, \tag{3c}$$

and

$$\left(\frac{\partial\left(\frac{\Delta G}{T}\right)}{\partial T}\right)_p = \frac{\Delta H}{T^2}. \tag{3d}$$

We won't derive any of these equations because the derivation is not particularly instructive. We will just show that the version of Equation 3a is true.

Start with the left-hand side of Equation 3a and show that it is equivalent to the definition of Gibbs free energy:

$$\left(\frac{\partial\left(\frac{G}{T}\right)}{\partial\left(\frac{1}{T}\right)}\right)_p = \left(\frac{\partial\left(\frac{G}{T}\right)}{\partial T}\right)_p \left(\frac{dT}{\partial\left(\frac{1}{T}\right)}\right)$$

$$= \left(\frac{\partial\left(\frac{G}{T}\right)}{\partial T}\right)_p \frac{1}{\frac{d\left(\frac{1}{T}\right)}{dT}}$$

$$= \left(\frac{\partial\left(\frac{G}{T}\right)}{\partial T}\right)_p \frac{1}{-\frac{1}{T^2}}$$

$$= -T^2 \left(\frac{\partial\left(\frac{G}{T}\right)}{\partial T}\right)_p$$

$$= -T^2 \left(+\frac{1}{T}\left(\frac{\partial G}{\partial T}\right)_p - \frac{1}{T^2}G\right)$$

$$= -T\left(\frac{\partial G}{\partial T}\right)_p + G$$

$$= -T(-S) + G$$

$$= G + TS$$

$$= H$$

We could easily substitute ΔG for G and end up with ΔH in the above sequence of equations.

So what is this good for? We will see that it is tremendously useful after we know the relationship between the equilibrium constant and $\Delta_r G^\circ/T$, but for now let's use the Gibbs-Helmholtz equation to calculate $\Delta_r G^\circ$ at a temperature, T_2 other than 25°C (which we will call T_1.

Set up Equation 3c for integration.

$$\left(\frac{\partial\left(\frac{\Delta_r G^\circ}{T}\right)}{\partial\left(\frac{1}{T}\right)}\right)_p = \Delta_r H^\circ, \tag{4}$$

$$d\left(\frac{\Delta_r G^\circ}{T}\right) = \Delta_r H^\circ d\left(\frac{1}{T}\right). \tag{5}$$

Equation 5 integrates to give,

$$\left(\frac{\Delta_r G^\circ}{T}\right)_{T_2} - \left(\frac{\Delta_r G^\circ}{T}\right)_{T_1} = \int_{T_1}^{T_2} \Delta_r H^\circ d\left(\frac{1}{T}\right). \tag{6}$$

To carry out the integration on the right-hand side of Equation 6 we would need to know how the heat of reaction changes with temperature. This information can be obtained if we know the heat capacities of the reactants and products as functions of temperature. However, usually the heat of reaction varies slowly with temperature so that it is a good approximation to regard the heat of reaction as constant.

Notice that Equation 5 is trivial to integrate if we make the approximation that $\Delta_r H^\circ$ is constant. With this approximation, Equation 5 integrates to,

$$\left(\frac{\Delta_r G^\circ}{T}\right)_{T_2} - \left(\frac{\Delta_r G^\circ}{T}\right)_{T_1} = \Delta_r H^\circ \int_{T_1}^{T_2} d\left(\frac{1}{T}\right), \tag{7}$$

or,

$$\left(\frac{\Delta_r G^\circ}{T}\right)_{T_2} - \left(\frac{\Delta_r G^\circ}{T}\right)_{T_1} + \Delta_r H^\circ\left(\frac{1}{T} - \frac{1}{T^1}\right). \tag{8}$$

We will apply Equation 8 to calculate ΔG for the freezing of super-cooled water at " 20°C. The process is,

$$H_2O(l, -20°C) \rightarrow H_2O(s, -20°C).$$

Do we know ΔG for this process at 0°C? Note that freezing water at its normal melting point is a reversible process so that the heat of fusion is a reversible heat at constant temperature. Therefore,

$$\Delta S_{fus} = \frac{\Delta H_{fus}}{T_{fus}}. \qquad (9)$$

Then

$$\Delta G_{fus} = \Delta H_{fus} - T_{fus}\Delta S_{fus}$$

$$\Delta H_{fus} - T_{fus}\frac{\Delta H_{fus}}{T_{fus}}$$

$= 0.$ (10a, b, c)

Equation 8, for our problem, becomes,

$$\left(\frac{\Delta G}{T}\right)_{T_2} = \left(\frac{\Delta G}{T}\right)_{T_2} + \Delta H\left(\frac{1}{T_2} - \frac{1}{T_1}\right)$$

$$= 0 - \Delta H_{fus}\left(\frac{1}{T_2} - \frac{1}{T_1}\right)$$

$$\left(\frac{\Delta G}{253.15K}\right) = -6008J\left(\frac{1}{253.15K} - \frac{1}{273.15K}\right)$$

Then ΔG is – 440 J which is what we would expect because freezing super-cooled water is a spontaneous (and irreversible) process at " 20°C.

We said above that $\Delta_r G^o$ tells us whether or not a reaction wants to go, but that the equilibrium constant, the ultimate arbiter of how strongly the reaction wants to go, depends on $\Delta_r G^o / T$. We will prove this statement later, but for now it might be interesting to see why that might be. Consider any process at constant temperature and pressure. We know that if T and p are constant then,

$$\Delta G = \Delta H - T\Delta S. \tag{11}$$

Let's divide Equation 11 by T to see what $\Delta G/T$ looks like,

$$\left(\frac{\Delta G}{T}\right) = \left(\frac{\Delta H}{T}\right) - \Delta S. \tag{12}$$

Note that ΔH and ΔS in Equation 12 refer to things happening in the system. Rewrite Equation 12 to indicate that this is true,

$$\left(\frac{\Delta G}{T}\right) = \left(\frac{\Delta H_{sys.}}{T}\right) - \Delta S_{sys}. \tag{13}$$

Now, ΔH_{sys} is heat absorbed by the system. Where did this heat come from? It had to come from the surroundings so.

$$\Delta H_{sun} = -\Delta H_{sys}$$

$$\Delta H_{sys} = -\Delta H_{sun}. \tag{14a, b}$$

Most likely our process is irreversibly, but that doesn't matter because H is a state function so that ΔH is independent of path. We can always find a reversibly path to change the enthalpy of the surroundings by an amount ΔH_{surr}. (Our process is at constant temperature and pressure so that ΔH in both cases is a heat at constant pressure.) If we add heat ΔH_{surr} to the surroundings isothermally and reversibly then the entropy change in the surroundings is

$$\Delta S_{sun} = \frac{\Delta H_{sun}}{T}$$

$$= -\frac{\Delta H_{sys}}{T}, \tag{15a, b}$$

or

$$\frac{\Delta H_{sys}}{T} = -\Delta S_{sun}. \tag{16}$$

Applying the result of Equation 16 to Equation 13 we find that,

$$\frac{\Delta G}{T} = -\Delta S_{sun} - \Delta S_{sys}$$

$$= -\Delta S_{universe}. \tag{17a, b}$$

Equation 17b tells us several things. It tells us that, in the final analysis, the ultimate driving force in nature is entropy, that is, the drive toward disorder. The system plus the surroundings is a closed isolated system so that the only spontaneous processes allowed are those which increase the entropy. Secondly, it explains why $\Delta G/T$ is more important than simply ΔG in determining how strongly spontaneous a process is. It is "$\Delta G/T$ which is related to the entropy change of the universe, not ΔG directly. Equation 17b also shows that the Gibbs free energy manages to include the entropy effects in the surroundings without ever telling us that it is doing so.

GIBBS FREE ENERGY AND PRESSURE, CHEMICAL POTENTIAL, FUGACITY

The Gibbs free energy depends on pressure as well as on temperature. The pressure dependence of the Gibbs free energy in a closed system is given by the combined first and second laws and the definition of Gibbs free energy as,

$$dG = sdT + Vdp \tag{1}$$

If we hold temperature constant and vary only the pressure we can prepare Equation 1 for integration from pressure p_1 to p_2 as follows:

$$dG = Vdp, \tag{2}$$

Then

$$G(p_2) - G(p_1) = \int_{p_1}^{p_2} Vdp \tag{3}$$

or

$$G(p_2) = G(p_1) + \int_{p_1}^{p_2} Vdp \tag{4}$$

Equation 4 is quite general and applies to all isotropic substance: solids, liquids, ideal gases, and real gases. We will apply it first to isotropic solids and liquids.

SOLIDS AND LIQUIDS

First level of approximation

We know that solids and liquids are not very compressible so, to a first approximation, we can regard the volume in Equation 4 as constant (as long as the range of pressure is not too large). Then the *V* in Equation 4 comes out of the integral and we can integrate easily to get,

$$G(p_2) = G(p_1) + V(p_2 - p_1) \tag{5}$$

Second level of approximation.

We know, however, that solids and liquids are slightly compressible and that we can define the isothermal compressibility as

$$k = -\frac{1}{V}\left(\frac{\partial V}{\partial p}\right)_T . \tag{6}$$

Our second level of approximation is to regard *k* as approximately constant. (In fact, *k* is not constant, but the variation with pressure is so small it can be ignored unless enormous pressures - megabars - are involved.) With *k* regarded as constant we can rearrange Equation 6 and integrate it to find an expression for *V* as a function of *p* (which can then be substituted into Equation 4 and integrated.) Rearrangement of Equation 6 looks like,

$$\frac{dV}{V} = -kdp \tag{7}$$

Integrate from p_1 to p_2 (and volume goes from V_1 to V_2) to get,

$$In\frac{V_2}{V_1} = -k(p_2 - p_1) \tag{8}$$

Take the antilog of both sides,

$$V_2 - V_{1e}^{-k(p_1 - p_2)} \tag{9}$$

In Equation 9 we can let p_1 be a constant and let p_2 range over the pressures of interest. There is no reason why we have to keep the subscript "2" on p_2 so change p_2 to just p. This gives us V as a function of p,

$$V_2 - V_{1e}^{-k(p - p_1) = V_{1e}^{kp_1 e^{-kp}}}. \tag{10}$$

On the far right of Equation 10 the constant parts have been separated from the variable part to make it easy to integrate. When this expression for V is plugged into Equation 4 only the e^{-kp} need stay inside the integral.

IDEAL GASES

Equation 4 is also valid for gases, only here we put in the value of V for an ideal gas.

$$V = \frac{nRT}{p} \tag{11}$$

With this substitution Equation 4 becomes,

$$G(p_2) = G(p_1) + \int_{p_1}^{p_2} \frac{nRT}{p} dp \tag{12}$$

which is an integral we have done many times. After integration Equation 12 becomes.

$$G(p_2) = G(p_1) + nRT\, In \frac{p_2}{p_1} \tag{13}$$

It is customary (and useful) to make several changes in Equation 13. We let p_2 range over the pressures of interest to us and call it just p, we let p_1 be some standard state pressure and call it p^o, and finally we divide through by the number of moles of gas, n. With these changes equation 13 is written,

$$\frac{G(p)}{n} = \frac{G(p^\circ)}{n} + RT\, In \frac{p}{p^\circ} \tag{14}$$

One more change: the quantity G/n turns out to be so important that it is given a special symbol and its own name. Strictly speaking G/n is just the Gibbs free energy per mole of substance, but the simplicity belies its importance. This quantity is called the chemical potential and it is given the symbol, μ . Our final version of what used to be Equation 13 is now,

$$\mu = \mu^{\circ} + RT\, In \frac{p}{p^{\circ}} \tag{15}$$

We have replaced $G(p^{o})/n$, the molar Gibbs free energy at the standard state pressure, with its chemical potential symbol, μ^{o}. In most cases we will set the standard state pressure equal to one atmosphere. It is not unusual to see Equation 15 written,

$$\mu = \mu^{\circ} + RT\, In\, p, \tag{16}$$

but when it is written like this we have to remember that there is an implied p^{o} = 1 atm dividing the p in the ln p, otherwise the argument of the log function would not be unitless.

NONIDEAL GASES

Equation 15 was derived assuming the gas is ideal. It does not apply to real gases or approximations to a real gas, like the van der Waals equation of state. If we know the equation of state we can go back to Equation 4 and make the appropriate modifications in our notation. That is, we divide Equation 4 by the number of moles, n, let p_1 equal the standard state pressure, p^{o} and note that V/n is the molar volume to get,

$$\mu = \mu^{\circ} + \int_{p}^{p} \cdot \overline{V}(p)\, dp' \tag{17}$$

(We have also let p_2 range over the pressures of interest and called it just plain p, which means that we have to change the dummy variable of integration from p to p'.) Equation 17 would provide the correct answer in numerical calculations, but it would wreak havoc in some of the later developments of thermodynamics, namely the equilibrium constant expression, as we will see later. G. N Lewis (the same Lewis of the Lewis dot structures and Lewis acid/base theory) proposed to preserve the form of Equation 15 by writing the chemical potential as,

$$\mu = \mu^{\circ} + RT\, In \frac{f(p)}{p^{\circ}} \tag{18}$$

This equation defines a quantity $f(p)$ called the fugacity. The fugacity has units of pressure and it is a function of pressure. It contains all the information on the nonideality of the gas. For an ideal gas the fugacity is the same as the pressure. Since all real gases become ideal in the limit as pressure goes to zero we must have.

$$\lim_{p \to 0} f(p) = p \tag{19}$$

We would like to have a way of calculating the fugacity from the equation of state for a gas. To do this go all the way back to Equation 2 and divide it by the number of moles, n,

$$d\frac{G}{n} = \frac{V}{n} dp, \tag{20}$$

or, in our new notation,

$$d\mu = \overline{V} dp. \tag{21}$$

We can get another expression for $d\mu$by taking the differential of Equation 18 (Remember that R, T, p^{o} and μ^{o} are constants.)

$$d\mu = RTd\, Inf. \tag{22}$$

The $d\mu$in Equations 21 and 22 must be the same, so we can set them equal to each other

$$RTd\, in\, f = \overline{V} dp. \tag{23}$$

Rearrange this to get,

$$d\,In\,f = \frac{\overline{V}}{RT}dp \tag{24}$$

We could integrate this equation directly, but that would sort of take us back to where we started from. Instead, we use a mathematical trick before we integrate it. Add and $\frac{dp}{p}$ subtract to the right hand side of Equation 24,

$$d\,In\,f = \frac{\overline{V}}{RT}dp - \frac{dp}{p} + \frac{dp}{p}. \tag{25}$$

You can see that we didn't really change anything. Regroup the terms in Equation 25,

$$d\,In\,f = \left(\frac{\overline{V}}{RT} - \frac{1}{p}\right)dp + \frac{dp}{p}. \tag{26}$$

Now integrate from p^{o} to p. (We will have to call our dummy variable of integration p' so as not to conflict with the limits of integration.) We get,

$$In\,f - In\,f_o = \int_{p_o}^{p}\left(\frac{\overline{V}(p)}{RT} - \frac{1}{p'}\right)dp' + In\,p - In\,p_o \tag{27}$$

where f_o is the fugacity at p_o. Move the to the right hand side,

$$In\,f = \int_{p_o}^{p}\left(\frac{\overline{V}(p)}{RT} - \frac{1}{p'}\right)dp' + In\,p\left(In\,f_o - In\,p_o\right) \tag{28}$$

Now we can take the limit where p_o goes to zero. We know that f_o goes to p_o as $p_o \to 0$ so the last two terms in parentheses on the right cancel each other in this limit. Equation 28 becomes,

$$In\,f = \int_{0}^{p}\left(\frac{\overline{V}(p)}{RT} - \frac{1}{p'}\right)dp' + In\,p \tag{29}$$

Equation 29 will suffice to calculate the fugacity, but it is customary to take the antilog of both sides to get,

$$f = pe^{\int_0^p \left(\frac{\overline{V}(p)}{RT} - \frac{1}{p'} \right) dp'} + pe^{\int_0^p \frac{1}{p'}(Z-1)dp'} \tag{30}$$

In the last segment of Equation 29, $z = \frac{p'\overline{V}}{RT}$, is the so-called compressibility factor. In either version of Equation of 29 it is easy to see that if the gas is ideal $f = p$. It requires an equation of state or experimental data to calculate a fugacity from either Equation 30 or Equaton 29. From the right-hand side of Equation 30 we can see that the second form of the virial expansion,

$$\frac{p\overline{V}}{RT} = 1 + B'p + C'p^2 + \ldots\ldots \tag{31}$$

would be the best choice for calculating fugacity

OPEN SYSTEMS

Up to now all of the systems we have been working with were closed systems. That is, no material moved in or out of the system. Now it is time to extend our discussion to open systems, in which material *can* move in or out of the system.

As usual we begin with the combined first and second laws of thermodynamics, only now we have to take into account that the internal energy, *U*, will depend on the number of moles of each component present. We write

$$U = U(S, V, n_1, n_2, n_3, \ldots), \tag{1}$$

so, instead of

$$dU = TdS - pdV \tag{2}$$

we must add the contribution of moving material in and out of the system. That is, we write,

$$dU = \left(\frac{\partial U}{\partial S} \right)_{V, n_1 n_2, \ldots} dS + \left(\frac{\partial U}{\partial V} \right)_{S, n_1 n_2, \ldots} dV$$

$$+\left(\frac{\partial U}{\partial n_1}\right)_{S,V,n_2n_3,...} dn_1 + \left(\frac{\partial U}{\partial n_2}\right)_{S,V,n_1n_3,...} dn_2 + ... \qquad (3)$$

The first two terms of Equation 3 must be the same as the combined first and second laws, Equation 2. The remaining terms in Equation 3 are new to us. They are clearly awkward to write so we invent a new symbol for the partial derivative,

$$\left(\frac{\partial U}{\partial n_1}\right)_{S,V,n_2n_3....} = \mu_1,$$

$$\left(\frac{\partial U}{\partial n_2}\right)_{S,V,n_2n_3....} = \mu_2, \qquad (4a, b)$$

and so·on.

(Notice that we are using the same symbol, μ, here that we used for G/n previously. That is not an accident, as we shall see. We called μ the chemical potential. In an open system with more than one component μ_i will·be the chemical potential of component i in the mixture.)

With this notation we can rewrite Equation 3 as,

$$dU = TdS - pdV + \mu_1 dn_1 + \mu_2 dn_2 + \mu_3 dn_{3+} ...; \qquad (5)$$

We can carry the terms accounting for the movement of material through to our other "energy" functions, H, A, and G.

From $H = U + pV$ we obtain,

$$dU = TdS - Vdp + \mu_1 dn_1 + \mu_2 dn_2 + \mu_3 dn_{3+} ...; \qquad (6a)$$

from $A = U - TS$ we obtain

$$dA = -SdT - pdV + \mu_1 dn_1 + \mu_2 dn_2 + \mu_3 dn_{3+} ...; \qquad (6b)$$

and from $G = U + pV$ " TS we obtain,

$$dG = -SdT + Vdp + \mu_1 dn_1 + \mu_2 dn_2 + \mu_3 dn_{3+} .... \qquad (6c)$$

From Equations 5 and 6a, b, and c we see that there are four different appearing ways to write the μ_i's. For example,

$$\mu_1 = \left(\frac{\partial U}{\partial n_1}\right)_{S,V,n_1 n_2,....} = \left(\frac{\partial H}{\partial n_1}\right)_{S,p,n_1 n_2,...} = \left(\frac{\partial A}{\partial n_1}\right)_{T,V,n_2 n_3,...}$$

$$= \left(\frac{\partial G}{\partial n_1}\right)_{T,V,n_1 n_2,...} \quad (7)$$

All of these definitions are equivalent, but the last one,

$$\mu_1 = \left(\frac{\partial G}{\partial n_1}\right)_{T,p,n_1 n_2,....}, \quad (8)$$

will be the most important one because it gives the change in Gibbs free energy that comes from adding or removing material at constant pressure and temperature. For a process a constant temperature and pressure dG becomes

$$dG = \mu_1 dn_1 + \mu_2 dn_2 + \mu_1 dn_1 + ...$$

$$\sum_i \mu_i dn_i \quad (9a, b)$$

Notice that the quantities, μ_i, are intensive variables. We can see this because they are an extensive variable, G divided by another extensive variable, n_i.

INTEGRATION OF d*U*

There are some unique features of the differential, dU, in Equation 5,

$$dU = TdS - pdV + \mu_1 dn_1 + \mu_2 dn_2 + \mu_3 dn_3 + ..., \quad (5)$$

which allow us to integrate it in an exceptionally simple manner. (This statement is not true for the differentials dH, dA, and dG.) In Equation 5 all the differentials (dU, dS, dV, dn_1, etc.) are differentials of extensive properties. That is, they all depend on the amount of material in the system. In addition, all the coefficients of these differentials (T, p, μ_1, etc.) are intensive properties

We can imagine integrating Equation 5 by starting out with the system in one container and transferring it to another container one differential drop at a time while holding all of the intensive variables constant. As we move the system from one container to another all the extensive quantities move to the new container in proportion to the size of the drop. Let's parameterize this process with a variable, x,

$$0 \le x \le 1,$$

where $x = 0$ means the system is entirely in its old container and $x = 1$ means the system has been entirely transferred to its new container. The "drop size" is given by the size of dx. Then,

$$dU = U\,dx,$$
$$dS = S\,dx,$$
$$dV = V\,dx,$$
$$dn_1 = n_1\,dx,$$
$$dn_2 = n_2\,dx, \qquad (10)$$

and so on.

We can now rewrite Equation 5 in terms of the differential, dx,

$$Udx = TSdx - pVdx + \mu_1 n_1 dx + \mu_2 n_2 dx + \mu_3 n_3 dx + ..., \qquad (11a, b)$$

In actual fact we don't have to integrate Equation 11a or 11b to get the desired result, just divide 11b by dx to get

$$Udx = (TS - pV + \mu_1 n_1 + \mu_2 n_2 + \mu_3 n_3 + ...)dx. \qquad (12)$$

If you insist on moving the system from one container to the other then we can do this by integrating Equation 11b from $x = 0$ to $x = 1$,

$$U\int_0^1 dx = TS - pV + \mu_1 n_1 + \mu_2 n_2 + \mu_3 n_3 + ...)\int_0^1 dx, \qquad (13)$$

which gives Equation 12 because the integral is equal to 1.

Equation 12 is called the integrated form of the combined first and second laws for an open system.

We can use Equation 12 to obtain "integrated forms" of H, A, and G,

$$H = U + pV = TS + \mu_1 n_1 + \mu_2 n_2 + \mu_3 n_3 + ...,$$

$$A = U - TS = -pV + \mu_1 n_1 + \mu_2 n_2 + \mu_3 n_3 + ..., \tag{14a, b}$$

but the most important one is

$$G = U - pV - TS = \mu_1 n_1 + \mu_2 n_2 + \mu_3 n_3 + ..., \tag{14c}$$

The latter equation is sometimes written

$$G = \sum_i \mu_i n_i, \tag{15}$$

where the summation is over all the components in the system.

THE GIBBS-DUHEM EQUATION

We now have (or can get) two different expressions for dG. One expression is Equation 6c,

$$dG = SdT + Vdp + \mu_1 dn_1 + \mu_2 dn_2 + \mu_3 dn_3 + ..., \tag{6c}$$

and the other can be obtained from Equation 15 as,

$$dG = \mu_1 dn_1 + \mu_2 dn_2 + \mu_3 dn_3 + ... n_1 d\mu_1 + n_2 d\mu_2 + n_3 d\mu_3 + ... \tag{16}$$

Setting Equations 6c and 16 equal to each other, and canceling terms that are the same on both sides we obtain,

$$-SdT + Vdp = n_1 d\mu_1 + n_2 d\mu_2 + n_3 d\mu_3 + ... \tag{17}$$

Equation 17 is called the Gibbs-Duhem equation. It tells us that the intensive variables in a system can not all be assigned values independently. That is, you can assign virtually any value you desire to all of the intensive variables but one, but the value of that last one will be predetermined by the values of the others.

The Gibbs-Duhem equation is most often used for processes at constant temperature and pressure, whence Equation 17 becomes,

$$n_1 d\mu_1 + n_2 d\mu_2 + n_3 d\mu_3 + ... = 0. \tag{18}$$

Consider a two-component system at constant T and p. Then,

$$n_1 d\mu_1 + n_2 d\mu_2 = 0, \tag{19a}$$

or

$$n_1 d\mu_1 = -n_2 d\mu_2, \tag{19b}$$

which tells us that if we change μ_2 then μ_1 responds with a change which depends on the ratio of the amounts of the two components.

In a one-component system the Gibbs-Duhem equation takes the form,

$$-SdT + Vdp = n_1 d\mu_1. \tag{20}$$

We rearrange this equation to give,

$$d\mu_1 = -\frac{S}{n_1}dT + \frac{V}{n_1}dp$$

$$\bar{S}dT + \bar{V}dp, \tag{21a, b}$$

which demonstrates that in a one component system the chemical potential is a function only of T and p.

COMMENT ON LEGENDRE TRANSFORMS

Recall that we get from Equation 12 to Equations 14a, b, and c by making Legendre transforms. Equation 14c shows that G is a natural function of T, p, and all the n_i's. Can we now make more Legendre transforms to obtain a new function which is a function of T, p, and all the μ_i's as in,

$$New\ Function = U - TS + pV - \mu_1 n_1 - \mu_2 n_2 - \mu_3 n_3 - \ldots ? \tag{22}$$

If we plug in Equation 12 for U here we get,

$$New\ Function = 0$$

This means that no such function exists. This is a consequence of the fact that the new function, if it existed, would be a function of only intensive variables (all independent) and we have seen from the Gibbs-Duhem equation that the intensive variables are not all independent.

MAXWELL'S RELATIONS REVISITED

Equations 5 and 6a, b, and c open up many new possibilities for Maxwell's relations. For example, from

Equation 6c we obtain two equations which may be useful later,

$$\left(\frac{\partial S}{\partial n_1}\right)_{T,p} = -\left(\frac{\partial \mu_1}{\partial T}\right)_{p,n_1}, \tag{23}$$

and

$$\left(\frac{\partial S}{\partial n_1}\right)_{T,p} = -\left(\frac{\partial \mu_1}{\partial T}\right)_{T,n_1}. \tag{24}$$

In equations 23 and 24 it is understood that all the other n's are also being held constant. Equations 23 and 24 will be useful when we start worrying about how the chemical potentials depend on temperature and pressure.

PHASE EQUILIBRIUM

We are all familiar with phase transitions where a substance transforms from one stable phase to another at an equilibrium temperature. For example, ice will be in equilibrium with liquid water at 273.15 K and 1 atmosphere pressure, or liquid water will be in equilibrium with water vapour at 373.15 K and one atmosphere. We will now take a look at the thermodynamic principles involved in phase transitions.

I will assume at the beginning that we know nothing about the conditions under which two phases can be in equilibrium. The only things we know are the criteria for equilibrium under certain conditions. That is, we know that

$$dS_{U,V} \geq 0 \tag{1}$$

for a closed isolated system (that is, the entropy seeks a maximum), and

$$dA_{T,V} \leq 0 \tag{2}$$

for a system as constant temperature and volume (that is,the Helmholtz free energy seeks a minimum), and

$$dG_{T,p} \leq 0 \tag{3}$$

for a system at constant temperature and pressure (the Gibbs free energy seeks a minimum).

We will do a series of three thought experiments under different sets of conditions and use the above criteria to tell us things about the temperature, pressure, and chemical potential of a system of two phases in equilibrium.

I. A closed isolated system.

Consider a closed isolated system consisting of two phases, phase α in equilibrium with phase β. We will call the temperature of the α phase T_α and the temperature of the β phase T_β.

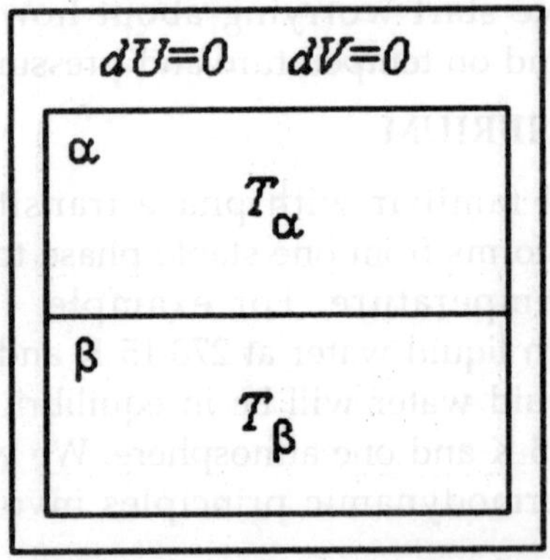

We do not know, yet, what is the relationship between T_α and T_β. We do know that in this system entropy seeks a maximum. That is,

$$dS_{U,V} \geq 0. \tag{4}$$

Let us transfer a small amount of heat, dq, reversibly from phase α to phase β. For definiteness we will set $dq > 0$. So

$$dq_\alpha = -dq$$

$$dq_\beta = +dq, \tag{5a,b}$$

and

$$dS_\alpha = \frac{dq_\alpha}{T_\alpha} = \frac{-dq_\alpha}{T_\alpha},$$

$$dS_\beta = \frac{dq_\beta}{T_\beta} = \frac{-dq_\beta}{T_\beta}, \tag{6a, b}$$

The total entropy change for the system is

$$dS = dS_\alpha + dS_\beta = \frac{dq_\alpha}{T_\alpha} + \frac{dq_\beta}{T_\beta} = \left(\frac{1}{T_\beta} - \frac{1}{T_\alpha}\right) dq \geq 0. \tag{7}$$

Since dq is positive by construction we conclude that

$$\left(\frac{1}{T_\beta} - \frac{1}{T_\alpha}\right) \geq 0,$$

$$\left(\frac{1}{T_\beta} \geq \frac{1}{T_\alpha}\right), \tag{8}$$

or

$$T_\beta \leq T_\alpha \tag{9}$$

If the system is not at equilibrium then $T_\beta < T_\alpha$, which makes sense because heat flows spontaneously from a higher temperature to a lower temperature.

If the system is at equilibrium then $T_\beta = T_\alpha = T$. There is only one temperature defined and there in no need to distinguish between the temperatures of the two phases.

II. A system at constant volume and temperature

Since we now know that at equilibrium there is only one temperature defined for the two phases let us now remove the system from isolation and place it in a heat bath at temperature, T. We will still hold the volume constant. Under conditions of constant temperature and volume we know that the criterion for equilibrium is that the Helmholtz free energy seeks a minimum. That is,

$dA_{T,V} \leq 0$. Our system now looks like,

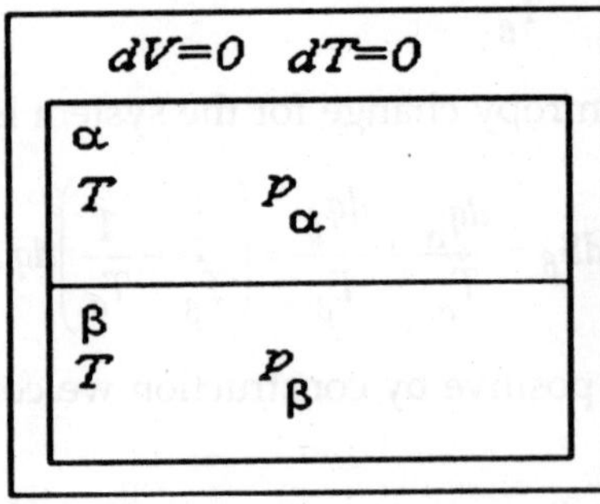

but we do not know the relationship between the pressures in the two phases.

Let us now transfer a small amount of volume, $dV > 0$, from phase α to phase β .

We know that *dA* = " *SdT* " *pdV* = " *pdV* for a constant temperature system. So,

$$dA_\alpha = p_\alpha dV_\alpha = -p_\alpha(-V),$$

$$dA_\beta = p_\beta dV_\beta = -p_\beta dV. \qquad (10)$$

The change in Helmholtz free energy for the entire system is,

$$dA = dA_\alpha + dA_\beta = p_\alpha dV - p_\beta dV = \left(p_\alpha - p_\beta\right) dV \leq 0. \qquad (11)$$

Since dV is positive we conclude that

$$p_\alpha - p_\beta \leq 0, \qquad (12)$$

or

$$p_\alpha \leq p_\beta.$$

If the system is not at equilibrium then $p_\alpha \leq p_\beta$, which makes sense since the β phase expanded at the expense of the α phase.

If the system is at equilibrium then $p_\alpha = p_\beta = p$, so that there is only one pressure defined for the system.

III. The system at constant pressure and temperature

Since we now know that at equilibrium both phases are at the same temperature and pressure, let us remove the constant volume restriction and place our system in a heat bath at temperature, T, and a pressure bath at pressure, p. (The atmosphere is a good example of a pressure bath at approximately one atmosphere pressure. You can make chambers to hold different pressures if you wish.) Our system looks like,

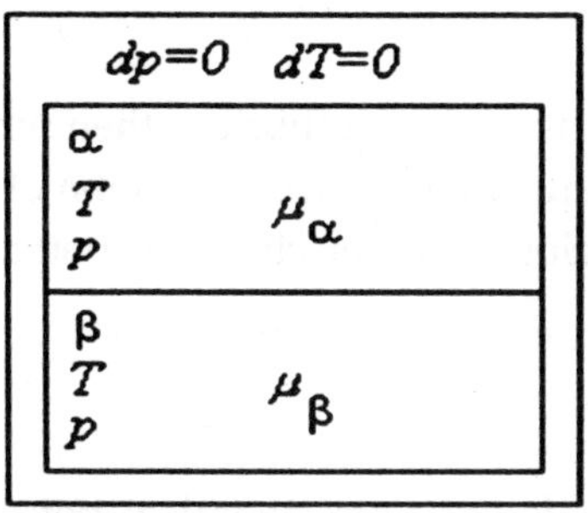

but we do not know the relationship between the chemical potentials in the two phases.

At constant temperature and pressure we know that

$$dG_{T,P} \le 0. \tag{13}$$

We also know that,

$$dG = -SdT + Vdp + \mu\, dn,$$

which at constant temperature and pressure becomes,

$$dG_{T,P} = \mu\, dn. \tag{14}$$

Let us now move a small amount of material, dn, from phase α to phase β (with $dn > 0$). Then,

$$dn_\alpha = -dn,$$

$$dn_\beta = +dn. \tag{15a, b}$$

and

$$dG_\alpha = \mu_\alpha\, dn_\alpha = -\mu_\alpha\, dn,$$

$$dG_\beta = \mu_\beta\, dn_\beta = -\mu_\beta\, dn, \tag{16a, b}$$

The change in Gibbs free energy for the entire system is then

$$dG = dG_\alpha + dG_\beta = \mu_\alpha\, dn + \mu_\beta\, dn = \left(\mu_\beta - \mu_\alpha\right) dn. \qquad (17)$$

Since dn is positive by construction we conclude that,

$$\mu_\beta - \mu_\alpha \le 0, \qquad (18)$$

or

$$\mu_\beta \le \mu_\alpha \qquad (19)$$

If the system is not in equilibrium then which makes sense since material wants to move from a region of higher chemical potential to a region of lower chemical potential.

If the system is in equilibrium then,

$$\mu_\beta = \mu_\alpha = \mu, \qquad (20)$$

so that there is only one chemical potential defined for the system.

Our conclusion, then, is that two phases in equilibrium must have the same temperature, pressure and chemical potential. The above sequence of derivations can easily be extended to include more phases and/or extended to include mixtures, where we would find that the temperature, pressure, and the chemical potential of each component must be the same in every phase. We will use these facts a little later in our derivation of the Gibbs phase rule.

ONE-COMPONENT PHASE DIAGRAMS

We know that most substances can exist in more than one phase. For example, solid, liquid, and gas. Some substances even have more than one solid phase. Carbon exists as graphite, diamond and a whole encyclopedia of forms related to C_{60}; tin exists as white tin and gray tin; and so on. A great amount of information on the phases of a substance can be summarized on one diagram called a "phase diagram."

Imagine that we take a sheet of graph paper and label the vertical axis (the ordinate) "*p*" for pressure, and the horizontal axis (the abscissa) "*T*" for temperature. Now examine the system at a particular temperature and pressure and label the corresponding point on the graph paper with a letter indicating the particular phase the system is in. That is, label the point with a "g," if the system is a gas, "l," for a liquid, "s," for a solid, and so on.

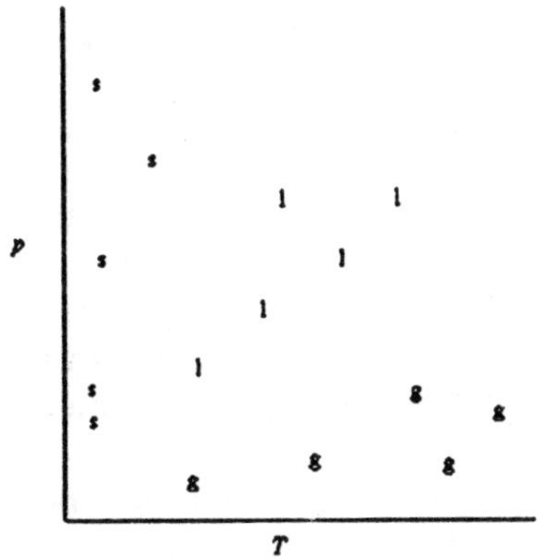

If we continued to do this at more and more points we would find that the graph paper would be separated into regions. In one of the regions all the points would be a "g" indicting that the substance is gas at those temperatures and pressures. In another region all the points would correspond to the substance being a liquid, and so on. In the limit where we considered an infinite number of points the regions would be separated from each other by curves.

Once we know the phase of a system in a particular region it is not usual to use so many g's and l's and s's. Most often we just label each region once.

This diagram is an example of a phase diagram. The line separating the solid and liquid regions is called the "melting curve," the line separating the liquid and gas regions is called the "vapourization curve," or the "vapour pressure curve," and the line separating the solid and gas regions is called the "sublimation curve." In the regions between the lines only one phase, the phase indicated, can exist at equilibrium.

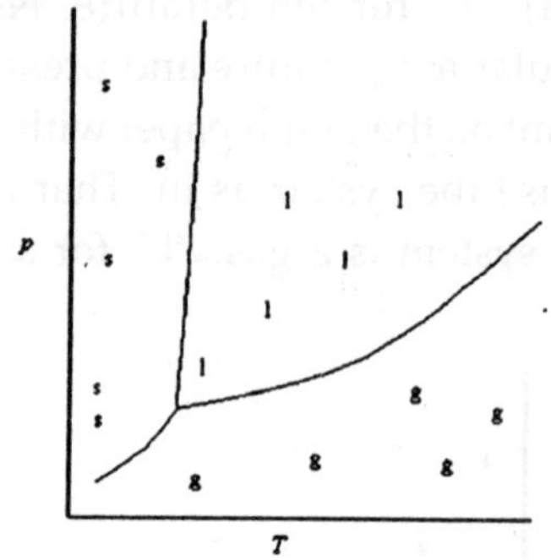

On a line we can have two phases in equilibrium with each other. The point where the three curves meet is called the "triple point." At the triple point we can have three phases in equilibrium. Notice that three phases can be at equilibrium with each other only at one point. This means that the triple point is "well defined" and reproducible. The triple point of water is used to define the Kelvin scale where, by definition, the triple point of water is 273.16 K. A substance that has more than one solid phase will have more than one triple point, but at the different triple points the three phases in equilibrium are never the same three.

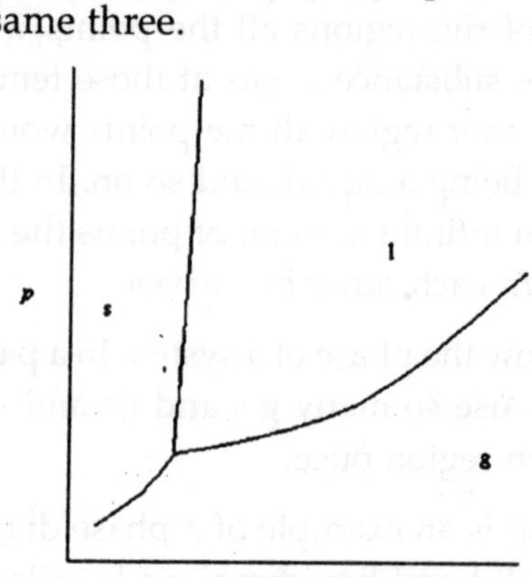

If the triple point has a presssure lower than one atmosphere the substance will have a normal melting point and a normal boiling point.

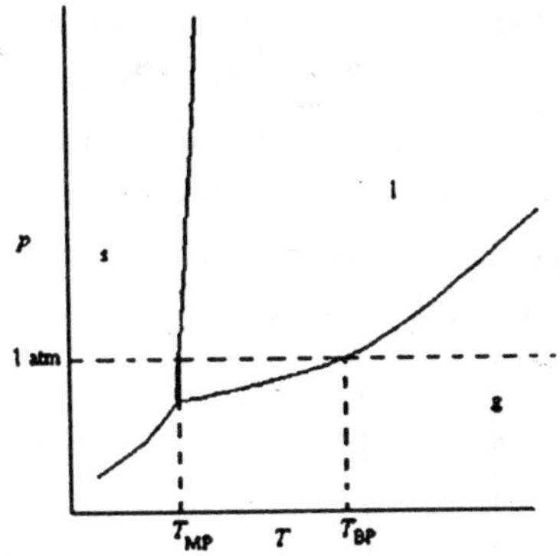

The normal melting point is the temperature at which the substance melts at one atmosphere pressure. (As you can see the melting point will be slightly different at other pressures.) The normal boiling point is the temperature where the substance boils at one atmosphere pressure. Another way to state this is that at the normal boiling point the vapour pressure of the substance equals one atmosphere.

As we learned in the early sections of the course, there is a point, called the critical point, "above" which the substance can not be converted to a liquid. That means that the vapour pressure curve must end at the critical point.

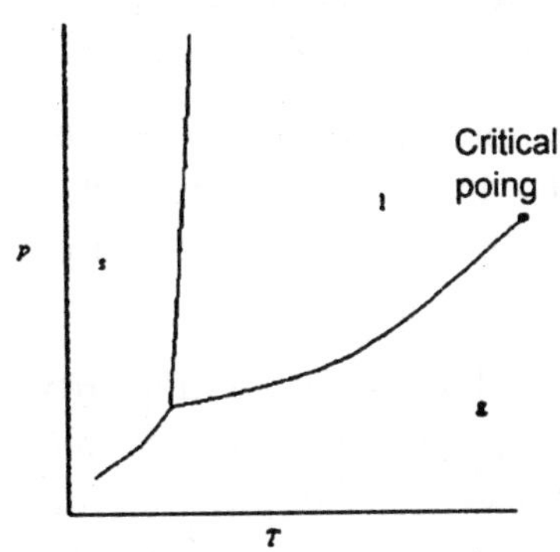

At pressures above the critical pressures it usually does not make sense to try to distinguish between whether the substance is a solid or a liquid. In these cases the substances is referred to simply as a "fluid."

CO_2 is a substance that does not have a normal melting point or boiling point because the substance is not liquid at one atmosphere pressure. (Its triple point pressure is above one atmosphere.) A phase diagram for CO_2 might look like the ones we have been showing, but the one-atmosphere line would be below the triple point.

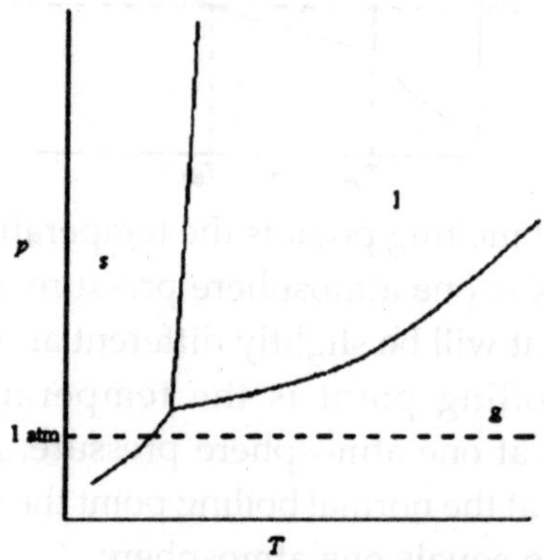

Solid CO_2 is called "dry ice" because it never melts at pressures near one atmosphere. At one atmosphere pressure the solid "sublimes" directly from the solid into the gas phase. In a CO_2 fire extinguisher you can hear the liquid "slosh around" inside the tank because liquid CO_2 does exist at pressures above the triple point pressure.

Water is an unusual substance because its melting point (273.15 K) is at a lower temperature than the triple point temperature (273.16 K). A schematic phase diagram for water might look something like,

Note that the slope of the melting curve is negative. The slope of the melting curve is positive for most substances. The negative slope for water is an important feature and we will discuss it in more detail after we have derived the Clapeyron equation. For now we note that the negative slope of the melting curve for water probably is a key feature in the ability of our planet to support life.

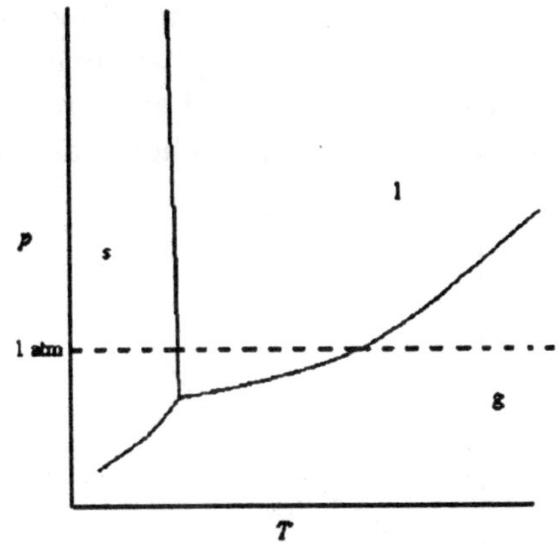

A completely labeled phase diagram for an ordinary substance might look like this,

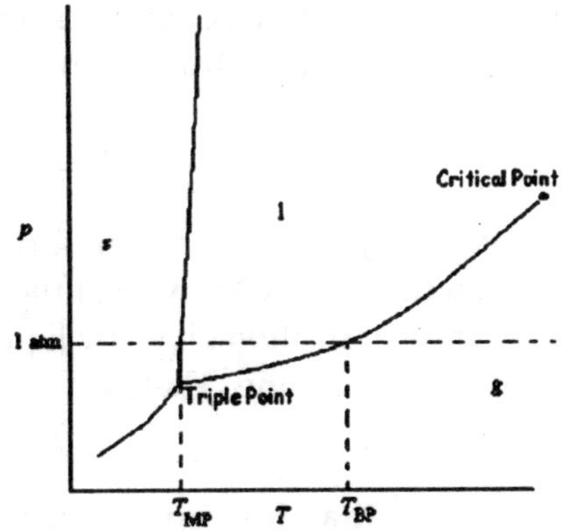

The dashed lines have been added simply to help locate the normal melting and boiling points of the substance and are not a necessary part of the phase diagram.

THE CLAPEYRON EQUATION

In a one-component phase diagram the lines separating the phases (for example, solid from liquid, liquid from vapour, and so on) are not arbitrary, they are determined by the principles of thermodynamics and equilibrium. Thermodynamics provides an equation for each line which gives pressure, p, as a function of temperature T, in other words an equation of the form $p = p(T)$. It is now our task to find the form of the function *p(T)*. The diagram below is a portion of a hypothetical phase diagram and the phase line on this diagram separates the phases α and β. In other words, the phases α and β are in equilibrium with each other at any and all points on the line.

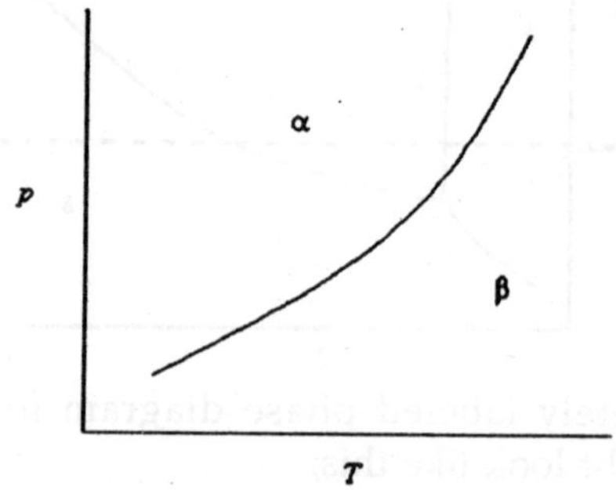

Phase transitions occur at constant temperature and pressure at the points on this line. Consider the phase transition,

$$\alpha \rightarrow \beta. \tag{1}$$

Since the α and β phases are in equilibrium with each other at any point on the line we know that the change in the Gibbs free energy for the transition given by Equation 1 is zero everywhere on the α-β phase line. That is,

$$\Delta G = 0. \tag{2}$$

We already know that for any process at constant temperature it must be true that

$$\Delta G = \Delta H - T\Delta S, \tag{3}$$

so that

$$\Delta S = \frac{\Delta H}{T} \tag{4}$$

for a phase transition.

The various quantities appearing in these equations are defined as usual by,

$$\Delta G = G_\beta - G_\alpha, \tag{5}$$

$$\Delta S = S_\beta - S_\alpha. \tag{6}$$

$$\Delta H = H_\beta - H_\alpha. \tag{7}$$

We will also define ΔV the same way,

$$\Delta V = V_\beta - V_\alpha. \tag{8}$$

As we have already said, $\Delta G = 0$ along the α-β phase line. That is, ΔG is constant along the line and it is being held constant at the particular value of zero. We want the slope of the line. Therefore, the derivative we want is, $\left(\frac{\partial p}{\partial T}\right)_{\Delta G.}$ We can write this derivative in terms of quantities we know using the cyclic rule (Euler's chain relation) as,

$$\left(\frac{\partial p}{\partial T}\right)_{\Delta G} = \frac{\left(\frac{\partial \Delta G}{\partial T}\right)_p}{\left(\frac{\partial \Delta G}{\partial p}\right)_T} \tag{9}$$

We know from $dG = -SdT + Vdp$, or $d\Delta G = -\Delta SdT + \Delta Vdp$, that

$$\left(\frac{\partial \Delta G}{\partial T}\right)_p = -\Delta S \tag{10}$$

and

$$\left(\frac{\partial \Delta G}{\partial p}\right)_T = -\Delta V \tag{11}$$

Using Equations10 and 11 in Equation 9 (the two minus signs cancel) we get,

$$\left(\frac{\partial p}{\partial T}\right)_{\Delta G} = \frac{\Delta S}{\Delta V} \tag{12}$$

Equation 12 is the Clapeyron equation. The Clapeyron equation is thermodynamically exact. It contains no approximations. There is another version of the Clapeyron equation which we obtain by inserting Equation 4 for ΔS into

Equation 12,

$$\left(\frac{\partial p}{\partial T}\right)_{\Delta G}=\frac{\Delta H}{T\Delta V} \tag{13}$$

Equation 13 is useful when we want to integrate dp to find p as a function of T. The integration of Equation 12 or 13 is easiest if we can make the approximation that either ΔS or ΔH is reasonably constant over the temperature range. However, ΔH usually varies more slowly with temperature than does ΔS, so that the approximation is better when we integrate 13. Prepare Equation 13 for integration,

$$dp=\frac{\Delta H}{T\Delta V}dT \tag{14}$$

To integrate Equation 14 we must know how ΔH and ΔV depend on temperature. We have already said that ΔH can be regarded as approximately constant, but what about ΔV? ΔV is certainly not even near constant if one of the components of the phase transition is a gas but if the transition is solid → solid or solid → liquid we can approximate ΔV as a constant also. Then Equation 14 integrates to,

$$p_2-p_1=\frac{\Delta H}{\Delta V}In\frac{T}{T_1} \tag{15}$$

If we let T_2 and p_2 range over the phase line and call them T and p, respectively, we can write pressure as a function of temperature for a solid → solid or solid → liquid phase transition line as follows,

$$p=p_1+\frac{\Delta H}{\Delta V}In\frac{T}{T_1} \tag{16}$$

THE CLAUSIUS-CLAPEYRON EQUATION

Equations 12 and 13 are exact and are valid for all types of phase transitions, solid-liquid, solid-gas, liquid-gas, and so on. However, for solid-gas and liquid-gas phase transitions we can make two approximations which will give us a very useful equation called the Calusius-Clapeyron equation. The first approximation comes from noting that the volume of a given amount of gas is much larger that the volume of an equivalent amount of solid or liquid. This being true, we can approximate ΔV by the volume of the gas. That is,

$$\Delta V = V_{gas} - V_{liquid}\ \mathrm{H}-\ V_{gas} \tag{17}$$

or

$$\Delta V = V_{gas} - V_{solid}\ \mathrm{H}-\ V_{gas}\ . \tag{18}$$

In the second approximation we replace V_{gas} by the volume the gas would have if it were ideal. That is, we set $V_{gas} = nRT/p$. With these two approximations Equation 13 becomes

$$\left(\frac{\partial p}{\partial T}\right)_{\Delta G} = \frac{\Delta H}{T\dfrac{nRT}{p}} = \frac{\Delta \overline{H}}{R}\frac{p}{T^2} \tag{19}$$

where the bar over the enthalpy, H, means that it is the enthalpy per mole. Equation 19 is the Clausius-Clapeyron equation. We should emphasize that Equation 19 is not exact (because it contains two approximations) and it only applies when one of the phases in the phase transition is a gas. Equation 19 can be prepared for integration is follows,

$$\frac{dp}{p} = \frac{\Delta \overline{H}}{R}\frac{dT}{T^2}. \tag{20}$$

Equation 20 can be integrated if we know how ΔH changes with temperature or, as we will now do, if we regard ΔH as approximately constant. With the ΔH regarded as constant Equation 20 integrates to

$$In\frac{p_2}{p_1} = \frac{\Delta \overline{H}}{R}\left(\frac{1}{T^2} - \frac{1}{T_1}\right) \tag{21}$$

Equation 21 is the integrated form of the Clausius-Clapeyron equation. If we want pressure as a function of temperature we can let the point (p_2, T_2) range over the phase line and call it simply (p, T) and then take the antilog of both sides,

$$p = p_1 e^{-\frac{\Delta H}{R}\left(\frac{1}{T}-\frac{1}{T_1}\right)} \tag{22}$$

Notice that if we know two points on the curve we can solve for ΔH, or if we know one point on the curve and ΔH we can solve for any other point or even get the entire curve. Equation 22 is very useful for estimating vapour pressures at various temperatures when ΔH and the vapour pressure at one temperature are known.

OTHER DETAILS AND INTERESTING STUFF

You can get fancy with Equations 19 and 20 if you want. For example, we know that ΔH for a phase transition is not really constant. If we wanted to take into account the temperature dependence of ΔH we need to know the heat capacities of the two phases, or rather the difference in the heat capacity for the two phases,

$$\Delta C_p = C_{p\beta} - C_{p\alpha}. \tag{23}$$

Then

$$\Delta \overline{H}(T) = \Delta \overline{H}(T_1) + \int_{T_1}^{T} \Delta \overline{C}_p dT' \tag{24}$$

Initially we regarded ΔH as constant. The next level of approximation would be to regard ΔC_p as constant so that,

$$\Delta \overline{H}(T) = \Delta \overline{H}(T_1) + \Delta \overline{C}_p (T - T_1) = \Delta \overline{H}(T_1) - \Delta \overline{C}_p T_1 + \Delta \overline{C}_p T \tag{25}$$

Using this expression for ΔH in Equation 20 we get

$$\frac{dp}{p} = \frac{\Delta \overline{H}(T_1) - T_1 \Delta \overline{C}_p dT}{R} \frac{dT}{T^2}, \tag{26}$$

$$\frac{dp}{p} = \frac{\Delta \overline{H}(T_1) - T_1 \Delta \overline{C}_p}{R} \frac{dT}{T^2} + \frac{\Delta \overline{C}_p}{R} \frac{dT}{T} \tag{27}$$

Equation 27 integrates to,

$$In \frac{p}{p_1} = \frac{\Delta \overline{H}(T_1) - T_1 \Delta \overline{C}_p}{R} \left(\frac{1}{T} - \frac{1}{T_1}\right) + \frac{\Delta \overline{C}_p}{R} In \frac{T}{T_1} \tag{28}$$

You can get even fancier. The next level of approximation would be to include the temperature dependence of the heat capacities, but we won't deal with that here.

Equation 12 does not depend on ΔG being zero. It only assumes that ΔG is constant for the α -β transition. Of course, if ΔG is not zero then equation 4 is not true which makes Equation 13 invalid. Further, if ΔG is not zero then either the α phase or the β phase is thermodynamically unstable. (We usually use the word *metastable* to describe this situation. Examples are supercooled liquids, superheated liquids, supercooled vapours, and some metastable solids - like diamond.) Nevertheless, one can visualize a family of curves running along side the phase diagram curve with ΔG constant at some value other than zero.

THE MELTING CURVE FOR WATER; VAPOUR PRESSURE

THE MELTING CURVE FOR WATER

We mentioned in our initial discussion of one-component phase diagrams that the slope of the melting curve for water was negative.

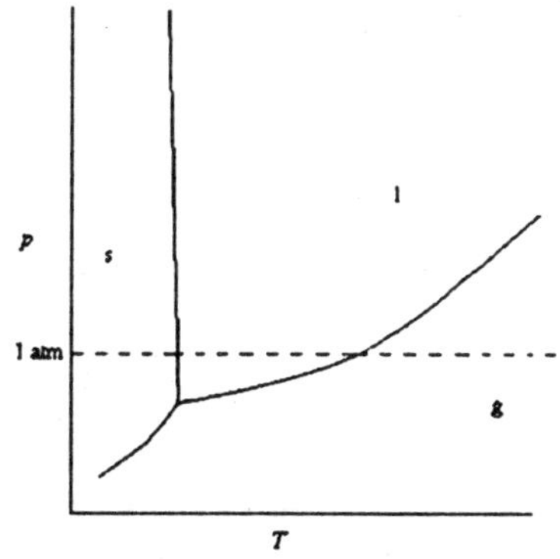

Water is one of just a few known substances with this feature (bismuth is another), the melting curve for most substances has a positive slope. Let us consider the thermodynamics consequences of this negative slope. Since we are talking about the melting curve we must use the Clapeyron equation,

$$\left(\frac{\partial P}{\partial T}\right)_{\Delta G} = \frac{\Delta S}{\Delta V}, \tag{1}$$

where, of course, the ΔS and ΔV refer to the melting process or the freezing process. For the purposes of this discussion let us consider the melting process. That is, we will assume that the phase change is

$H_2O(s)$-$H_2O(l)$.

A negative slope on the phase diagram means that the derivative in Equation 1 is negative. This in turn requires that,

$$\left(\frac{\partial P}{\partial T}\right)_{\Delta G} = \frac{\Delta S}{\Delta V} < 0. \tag{2}$$

We know that ΔS for the melting process is positive because the system is going from the highly ordered crystalline solid state to the more disordered liquid state (and we can calculate it from $\Delta S = \Delta H_{fusion}/T_{mp}$). Therefore, the only way the derivative in Equation 2 can be negative is if ΔV is negative. This means that the molar volume of the liquid is *smaller* than the molar volume of the solid at the melting point,

$$\overline{V}_l - \overline{V}_S < 0,$$

$$\overline{V}_l < \overline{V}_S \tag{3a, b}$$

Taking the reciprocal of Equation 3b we find that,

$$\frac{1}{\overline{V}_l} > \frac{1}{\overline{V}_S} \tag{4}$$

and, since the density is proportional to the reciprocal of the volume, this means that the density of the liquid is greater than the density of the solid (at the melting point).

This result is counterintuitive. We expect that materials will expand on melting and most materials do expand on melting. However, water is unusual in this respect and contracts on melting. The visible consequence of this fact is that ice floats on liquid water. We are so accustomed to this that we do not even notice that it is unusual. However, if you have worked in the laboratory with other solid/liquid systems you will be aware that most often the solid sinks in the melt.

You can argue that the slope of the melting curve of water makes a crucial contribution to the ability of our planet, Earth, to support life. If the slope were positive then ice would sink in water. In cold climates ice would form in winter and sink to the bottom. This ice would be insulated from the summer warmth by the layer of water above it and would only partially melt. The next winter more ice would accumulate on the bottom. Eventually, all rivers and lakes and possibly a large portion of the oceans would be frozen from the bottom up with only a shallow layer of liquid water being formed on the surface during the summer. There would likely be no animal or plant life in these frozen lakes and rivers.

VAPOUR PRESSURE

We mentioned in our discussion of one-component phase diagrams that the vapourization curve is sometimes referred to as the vapour pressure curve. This is because this curve - along with the sublimation curve - gives a plot of the vapour pressure of the substance versus temperature.

The vapour pressure is a measure of the ability of molecules to escape from the surface of a solid or liquid. The following discussion uses ideas that are outside the scope of thermodynamics, but they may be useful in calibrating our intuition. A solid or liquid is held together by intermolecular attractive forces. However, the molecules have kinetic energy (energy of motion) which we will later see is proportional, on the average, to the Kelvin temperature. There is a competition. The attractive forces are trying to keep the molecules in the solid or liquid and the kinetic energy is trying to take them out. The vapour pressure is the net result of this competition.

As more and more molecules build up in the space above the solid or liquid more of them will collide back with the surface and some of them will stick. At a fixed temperature the system will reach a state of "dynamic equilibrium," where just as many molecules escape from the surface as collide back and stick. At this point the pressure exerted by the molecules in the space above the solid or liquid is, by definition, the vapour pressure of the solid or liquid.

As the temperature is increased the average kinetic energy increases so more molecules are able to overcome the intermolecular forces and the vapour pressure increases.

INCREASING THE VAPOUR PRESSURE BY THE APPLICATION OF AN EXTERNAL PRESSURE

One can also increase the vapour pressure of a solid or liquid by applying an external pressure. This effect is not as easy to understand qualitatively as the effect of temperature, but it is easy to calculate thermodynamically. Recall that for solids and liquids, which are not very compressible over relatively small pressure ranges, the Gibbs free energy change is given by,

$$\Delta G \approx V(p - p_o). \tag{5}$$

In words, Equation 5 says that G increases when we "squeeze" a solid or a liquid. (The Gibbs free energy of a gas also increases when you "squeeze it, but gases are quite compressible so that Equation 5 does not hold for gases.) We would expect that if the Gibbs free energy of a solid or liquid increases its vapour pressure would also increase. We shall see if this is true. Rewrite Equation 5 as follows,

$$\Delta G = G_p - G_{p_o} = V(p - p_o). \tag{6}$$

We will rearrange Equation 6 and then divide it by the number of moles of substance, n.

$$G_p = G_{po} + V(p - p_o),$$

$$(G_p = G_{po} + V(p - p_o))\frac{1}{n},$$

$$\frac{G_p}{n} = \frac{Gp_o}{n} + \frac{V}{n}(p - p_o),$$

$$\mu_p = \mu_{p_o} + \overline{V}(p - p_o). \tag{7a, b, c, d}$$

In Equation 7d we have used the fact that the Gibbs free energy per mole is called the chemical potential and is symbolized by a lower case Greek "mu."

Let us consider two systems, all at temperature, T. The first system is a closed container which contains only, say, liquid water and water vapour in equilibrium. Since the system is in equilibrium the chemical potential of the liquid water and water vapour must be equal. Using the approximation that the vapour is an ideal gas we can write,

$$\mu_l^o = \mu_g,$$

$$\mu_l^o = \mu_g^o, + RT\, In \frac{p_o}{1\,atm}. \qquad (8a, b)$$

(In Equation 8b we have used the expression for the variation of the chemical potential of an ideal gas with pressure.)

Our second system consists of a similar container of liquid water and water vapour, but also contains an inert gas, say air, at a pressure, P. It is the air that is squeezing the liquid and providing the "external pressure." We will call this system the "prime" system. In this second system the chemical potentials of the liquid water and water vapour must still be equal, but they are not equal to the same values as in the first system. In the "prime" system we have,

$$\mu_1' = \mu_g', \qquad (9)$$

but

$$\mu_i' = \mu_i^o + \overline{V}(P-0) = \mu^\circ + \overline{V}P \qquad (10)$$

and

$$\mu_g' = \mu_g^o + RT\, In \frac{p'}{1\,atm}. \qquad (11)$$

So we have two equations equating chemical potentials in the gas and liquid phases, one for the first system and one for the second system.

$$\mu_i^\circ + \overline{V}P = \mu_g^\circ + RT\, In \frac{p'}{1atm}, \tag{12}$$

and

$$\mu_i^\circ = \mu_g^\circ + +RT\, In \frac{p_\circ}{1atm}. \tag{12b}$$

When we subtract Equation 8b from Equation 12 we eliminate all the standard state chemical potentials and are left with

$$\overline{V}P = RT\, In \frac{p'}{1atm} - RT\, In \frac{p_\circ}{1atm}$$

$$= RT\, In \frac{p'}{p_\circ}, \tag{13a, b}$$

so that

$$p' = p_\circ e^{\frac{\overline{V}p}{RT}}. \tag{14}$$

Equation 14 tells us that the vapour pressure of the liquid, in this case water, increases when we apply an external pressure. For water the vapour pressure increases by 0.074 per cent at 1.00 atm and 7.6 per cent at 100 atm.

MIXTURES; PARTIAL MOLAR QUANTITIES; IDEAL SOLUTIONS

MIXTURES

We have seen that the combined first- and second laws for an open system is written,

$$dU = TdS - pdV + \mu_1 dn_1 + \mu_2 dn_2 + ..., \tag{1}$$

Where the initial definitions of the μ_i were given by,

$$\mu_1 = \left(\frac{\partial U}{\partial n_1}\right)_{S,V,n_2,...} \tag{2}$$

An equivalent, and more easily understood, definition of the μ_i was given by,

$$\mu_1 = \left(\frac{\partial G}{\partial n_1}\right)_{T,p,n_2,\ldots} . \tag{3}$$

μ_i is called the "chemical potential" and it can be described in words as the Gibbs free energy per mole of substance. Another name for this quantity is "partial molar Gibbs free energy." We found, using the integrated form of Equation 1 and the definition of the Gibbs free energy, that the Gibbs free energy could be written as,

$$G = \mu_1 n_1 + \mu_2 n + \ldots \tag{4}$$

Equation 4 is an exact thermodynamic equation and contains no approximations.

Notice in Equation 3 that μ_i is intensive, it is given by the extensive G divided by the extensive n_i. Also, the variables that are being held constant are temperature and pressure along with the number of moles of the other components of the system. We can define an equation similar to Equation 3 for any extensive property. The partial molar Gibbs free energy is the most important of these because it provides a measure of the "driving force" for chemical processes. The next most important of these quantities is the partial molar volume,

$$\overline{V}_i = \left(\frac{\partial V}{\partial n_i}\right)_{T,p,n,\ldots} . \tag{5}$$

(We will use the bar over the symbol to indicate partial molar quantities. Some texts use a subscript, m, as in V_{mi} to indicate them.) The partial molar volume in Equation 5 can be thought of in several ways. It is the incremental volume obtained by adding a small amount of component *i* to the mixture while holding the temperature, pressure, and the number of moles of all the other components constant divided by the number of moles of component *i*. Another way to look at it is to say that it is the incremental volume obtained by adding one mole of component i to an infinite sample of the mixture. The partial molar volume is not necessarily the same as the volume of one mole of the pure component. (The intermolecular interaction between molecules of one component and molecules of other components may be different than the interaction of molecules of a component with other molecules of the same component.)

Let us regard V as a function of temperature, pressure and composition:

$$V = V(T, p, n_1 n_2 ...).$$

Then

$$dV = \left(\frac{\partial V}{\partial T}\right)_{p,n} dT + \left(\frac{\partial T}{\partial n}\right)_{T,n} dp + \left(\frac{\partial V}{\partial n_1}\right)_{T,p,n_2 ..} dn_1 + \left(\frac{\partial V}{\partial n_2}\right)_{T,p,n_1} dn_2 + ...$$

$$\left(\frac{\partial V}{\partial T}\right)_{p,n} dT + \left(\frac{\partial T}{\partial n}\right)_{T,n} dp + \overline{V_1} dn_1 + \overline{V_2} dn_2 + ... \quad (6a, b)$$

If we hold T and p constant we get

$$dp + \overline{V_1} dn_1 + \overline{V_2} dn_2 + ..., \quad (7)$$

which can be integrated (similar to the way we integrated Equation 1, above) to give,

$$V = n_1 \overline{V}_1 + n_2 \overline{V}_2 + ...,$$

$$= \sum_i n_i \overline{V}_i \quad (8a, b)$$

The volumes given in Equations 7, 8a, and 8b are the partial molar volumes which we have already said are not necessarily equal to the molar volumes of the pure components. Thus, Equations 8a and b tell us that volumes may not be additive. That is, if we were to mix one liter of pure ethanol with one liter of pure water the final volume of the mixture would not likely be two liters. That is because water molecules interact with ethanol molecules differently than they interact with other water molecules.

As we have said above, we can define a partial molar quantity from any extensive variable. Thus we can define the partial molar entropy as

$$\overline{S}_i = \left(\frac{\partial S}{\partial n_i} \right)_{T,p,n_j}, \tag{9}$$

or the partial molar enthalpy as,

$$\overline{H}_i = \left(\frac{\partial H}{\partial n_i} \right)_{T,p,n_j}, \tag{10}$$

It is also true that

$$S = n_1 \overline{S}_1 + n_2 \overline{S}_2 + \dots, \tag{11}$$

and so on.

How to Measure Partial Molar Volumes

There are several ways that partial molar volumes can be measured. One way is to begin with one mole of a compound, call it component 1, add a small amount of component 2 and measure the volume, add a little more of component 2 and measure the volume again. Keep doing this until the desired concentration range has been covered. Then fit the volume data to a curve, for example, of the form,

$$V(n_1 = 1, n_2) = \overline{V_1^\circ} + an_2 + bn_2^{3/2} + cn_2^2 + \dots \tag{12}$$

(Why would we not want to include that 1/2 power of n_2 in Equation 12? People use whatever powers of n_2 they need fit the data.) The constants, *a, b, c,* etc are obtained from the curve fitting and the first term is the molar volume of pure component 1. Then the partial molar volume of component 2 can be obtained by direct differentiation,

$$\left(\frac{\partial V}{\partial n_2}\right)_{T,p,n_1} = \overline{V_2}$$

$$= 0 + a + \frac{3}{2}bn_2^{1/2} + 2cn_2 + ... \qquad (13a, b)$$

The partial molar volume of component 1 can be obtained from,

$$V = n_1\overline{V}_1 + n_2\overline{V}_2$$

$$= \overline{V}_1 + n_2\overline{V}_2, \qquad (14a, b)$$

or

$$\overline{V_1} = \frac{V - n_2\overline{V}_2}{n_2}$$

$$= V - n_2\overline{V}_2. \qquad (15a, b)$$

(The last step is because, in this case we have set $n_1 = 1$.)

IDEAL SOLUTIONS

We will define an ideal solution as a solution for which the chemical potential of each component is given by,

$$\mu_i = \mu_i^* + RT\, In\, X_i, \qquad (16)$$

where μ_i^* is the chemical potential of pure component *i*, and X_i is the mole fraction of component *i* in the solution.

(Many texts define an ideal solution as a solution which obeys Raoult's law over the full range of composition,

$$p_i = X_i p_i^*, \qquad (17)$$

where p_i^* is the vapour pressure of pure component *i*.)

We will now prove that an ideal solution obeys Raoult's law (using our definition of an ideal solution).

Consider a solution of two components where the mole fraction of component 1 is X_1. We know that the chemical potential of component 1 must be the same in the solution as in the vapour in equilibrium with the solution. That is,

$$\mu_{1l} = \mu_{1g}, \tag{18}$$

but the solution is ideal so,

$$\mu_{1l} = \mu^*_{1/l} + RT\, In\, X_1. \tag{18a}$$

Also, we can approximate the vapour as an ideal gas so,

$$\mu_{1g} = \mu^\circ_{1g} + RT\, In \frac{p_1}{1atm}, \tag{19}$$

where p_1 is the vapour pressure (partial pressure) of component 1 above the solution. Combining Equations 16, 18, and 19 we get,

$$\mu^*_{1/l} + RT\, In\, X_1 = \mu^\circ_{1g} + RT\, In \frac{p_1}{1atm}. \tag{20}$$

Equation 20 doesn't help us very much all by itself. However we have some more information. We know that for the pure component 1 we have $X_1 = 1$, and we know that the pressure of component 1 vapour in equilibrium with the liquid is just the vapour pressure of the pure liquid, p_1^*, so that,

$$\mu^*_{1/l} = \mu^\circ_{1g} + RT\, In \frac{p^*_1}{1atm}. \tag{21}$$

Let us now subtract Equation 21 from Equation 20 to get

$$RT\, In\, X_1 = RT\, In \frac{p_1}{1atm} - RT\, In \frac{p^*_1}{1atm}, \tag{22}$$

from which we conclude that,

$$In\, X_1 = In \frac{p_1}{p^*},$$

$$X_1 = \frac{p_1}{p^*}, \tag{23a, b}$$

or

$$p_1 = X_1 p^*, \tag{17}$$

which is Raoult's law.

EXAMPLE CALCULATION USING RAOULT'S LAW

Benzene and toluene form a solution which is very nearly ideal. Consider a mixture of benzene (Bz) and toluene (Tol) at 60° C. At 60° C the vapour pressures of pure benzene and pure toluene are 385 Torr and 139 Torr, respectively. What are the vapour pressures of benzene and toluene in a mixture with $X_{Bz} = 0.400$, and $X_{Tol} = 0.600$, and what is the composition of the vapour in equilibrium with this solution?

Use Raoult's law to find the vapour pressures of the two species,

$$P_{Bz} = X_{Bz} P_{Bz}^* = 0.400 \times 385\, Torr = 154\, Torr,$$

$$P_{Total} = X_{Tol} P_{Tol}^* = 0.600 \times 139\, Torr = 83.3\, Torr.$$

The total pressure is the sum of these two individual pressures, 237 Torr.

The composition of the vapour phase is obtained from the vapour pressures and Dalton's law of partial pressures,

$$X_{Bz} = \frac{P_{Bz}}{P_{Tot}} = \frac{154}{237} = 0.649,$$

$$X_{Tol} = \frac{P_{Tol}}{P_{Tol}} = \frac{83.3}{237} = 0.351.$$

Notice that the composition of the vapour is not the same as the composition of the liquid, the vapour phase is much richer in the more volatile compound, benzene. This fact will be important when we discuss vapour pressure diagrams and two-component phase diagrams.

PROPERTIES OF IDEAL SOLUTIONS

Given the chemical potentials for the components of an ideal solution we can calculate a number of properties of ideal solutions. For example, the Gibbs free energy of mixing is the easiest to calculate,

$$\Delta G_{mix} = G_{mixed} - G_{unmixed},$$

$$n_1\mu_1 + n_2\mu_2 + ... - n_1\mu_1^* + n_2\mu_2^* - ....$$

$$n_1\mu_1^* + n_2\mu_2^* - ...$$

$$RT\left(n_1 \, In X_1 + n_2 \, In X_2 + ...\right). \qquad (24a, b, c, d)$$

To bring this equation into the usual form multiply and divide by the total number of moles, n, and bring the $1/n$ inside the parentheses to convert the number of moles of each component into a mole fraction,

$$\Delta G_{mix=nRT(X_1 In X_1 + X_2 In X_2 + ...)}. \qquad (25)$$

Notice that the Gibbs free energy of mixing is negative, as one would expect for a spontaneous process at constant temperature and pressure.

We can also calculate other properties of mixing ideal solutions. The entropy of mixing is,

$$\Delta S_{mix} = -\left(\frac{\partial \Delta G_{mix}}{\partial T}\right)_{p_i n_j}$$

$$= -nR\left(X_1 \, In X_1 + X_2 \, In X_2 + ...\right), \qquad (26)$$

which is the same as the entropy of mixing for ideal gases.

The volume change on mixing can be found from,

$$\Delta V_{mix} = \left(\frac{\partial \Delta G_{mix}}{\partial p}\right) T, n_i = 0, \qquad (27)$$

so that volumes are additive for an ideal solution. That is, if we were to mix one liter of benzene and one liter of toluene the final volume of the solution would be two liters.

We can also determine whether or not there is any heat of reaction.

$$\Delta G_{mix} = \Delta G_{mix} + T\Delta S_{mix}$$
$$= nRT(X_1 InX_1 + X_2 InX_2 + ...)$$
$$= +T(-nR)(X_1 InX_1 + X_2 InX_2)$$
$$= 0. \quad (28)$$

That is, there is no heat involved in the mixing of ideal solutions. If we mix several components and the mixture gets hot or cold we can be sure that the solution formed was not ideal.

ACTIVITY AND ACTIVITY COEFFICIENTS

We have now seen and used several expressions for the chemical potential of a substance or component in a mixture. For a one-component ideal gas we had

$$\mu = \mu^\circ + RTIn\frac{p}{p^\circ}, \quad (1)$$

where μ^o is the chemical potential when $p = p^o$ and p^o is usually one atmosphere.

For a mixture of ideal gases it can be shown that for each component, *i*, the chemical potential is given by,

$$\mu_i = \mu_i{}^\circ + RTIn\frac{p_i}{p^\circ}, \quad (2)$$

For a nonideal gas we used the fugacity, *f*, instead of the pressure and the chemical potential for one component is,

$$\mu = \mu^\circ + RTIn\frac{f(p)}{p^\circ}. \quad (3)$$

For a mixture of nonideal gases it can be shown that,

$$\mu_i = \mu_i{}^\circ + RTIn\frac{f_i}{p^\circ}, \quad (4)$$

only now the fugacity of component *i* is a function of the pressures of all the gases in the mixture,

$$f_i = f_i(p_1, p_2, p_3, ...). \quad (5)$$

For ideal solutions we found that,

$$\mu_i = \mu_i^\circ + RT In X_i. \tag{6}$$

These expressions for chemical potential all have the form of a reference or standard state chemical potential plus RT times the logarithm of something related to pressure or concentration. This form turns out to be very important, so important that G. N. Lewis used it to give the most general case of chemical potential as

$$\mu_i = \mu_i^\circ + RT In a_i. \tag{7}$$

The quantity, a_i, is called the "activity" of component i and Equation 7 should be regarded as the definition of activity. Notice that the activity has no units.

All of the special cases we have been considering so far can be reconciled to this definition of activity. Thus, for an ideal gas mixture,

$$a_i = \frac{p_i}{p^\circ};$$

for a nonideal gas mixture,

$$a_i = \frac{f_i}{p^\circ};$$

for an ideal solution,

$$a_i = X_i,$$

and so on. In a nonideal solution we would have to just write Equation 7 again,

$$\mu_i = \mu_i^\circ + RT In a_i. \tag{7}$$

In Equation 7 all the nonidealities of the solution are absorbed into the activity. We will see a more convenient way to write this below under the heading "activity coefficient."

We can find the activity of a component of a nonideal solution from measurements of the vapour pressure of that component in the vapour in equilibrium with the solution. We know that the chemical potential of a component must be the same in the vapour as in the liquid. that is, from Equations 2 and 7 we obtain,

$$\mu_{il} = \mu_{ig} \tag{8a, b}$$

$$\mu^*_{il} = RTIna_i = \mu^\circ_{ig} + RTIn\frac{p}{p^\circ},$$

but for pure component i, we must have,

$$\mu^*_{il} = \mu^\circ_{ig} + RTIn\frac{p^*_i}{p^\circ}. \tag{9}$$

(Note: When dealing with liquid solutions it is customary to write the chemical potential of the pure liquid as μ^*_{il} instead of the usual μ^o_{il}, which means that the standard state for a liquid is the pure liquid itself. p_i^* is the vapour pressure of the pure liquid.)

Subtracting Equation 9 from Equation 8b yields,

$$RTIna_i = RTIn\frac{p_i}{p^\circ} - RTIn\frac{p^*_i}{p^\circ}$$

$$RTIn\frac{p_i}{p^*_i}, \tag{10a, b}$$

so that

$$a_i = \frac{p_i}{p^*_i}. \tag{11}$$

If the solution were ideal p_i would be given by Raoult's law and the activity would be just the mole fraction.

ACTIVITY COEFFICIENT

Sometimes it is convenient to write the activity as the product of an ideal part times a nonideality correction part. For example, in a real solution we might write,

$$a_i = X_i\gamma_i, \tag{12}$$

or, in a real gas we would write,

$$f_i = p_i \gamma_i, \tag{13}$$

so that

$$a_i = \frac{f_i}{p^\circ} = \frac{p_i \gamma_i}{p^\circ} = \frac{p_i}{p^\circ} \gamma_i. \tag{14}$$

In the latter case we already know that

$$f(p) = p e^{\int_0^p \frac{(Z-1)}{p'} dp'}, \tag{15}$$

so that

$$\gamma = e^{\int_0^p \frac{(Z-1)}{p'} dp'}. \tag{16}$$

In the case of the nonideal gas $\gamma \to 1$ as $p \to 0$. In the case of a nonideal solution $\gamma_i \to 1$ as $X_i \to 1$. From Equations 11 and 12 we see that,

$$a_i = \frac{p_i}{p_i^*} = X_i \gamma_i, \tag{17}$$

from which we conclude that for a nonideal solution,

$$\gamma_i = \frac{p_i}{X_i p_i^*} = \frac{\textit{actual vapor pressure}}{\textit{Raoult's law vapor pressure}}. \tag{18}$$

In general, the activity coefficient is a unitless parameter that contains all of the nonideality of a system.

VAPOUR PRESSURE DIAGRAMS AND BOILING DIAGRAMS

We are now ready to begin talking about phase diagrams involving two components. Our first few phase diagrams will involve only the liquid and gas (or vapour) phases. Later we will discuss two-component phase diagrams involving liquids and solids.

Consider a solution of two liquids, say benzene and toluene held at a constant temperature. Benzene and toluene form very nearly ideal solutions so we shall regard the solution as ideal. We know that an ideal solution obeys Raoult's law, so for each component we must have,

$$p_i = X_i p_i^*, \tag{1}$$

where p_i is the partial pressure of component i above the liquid mixture (or the vapour pressure of i in the mixture), X_i is the mole fraction of component i in the liquid, and p_i^* is the vapour pressure of pure component i.

If we now hold temperature constant and plot the vapour pressure of a benzene-toluene solution as a function of composition we would get a graph that looked something like this,

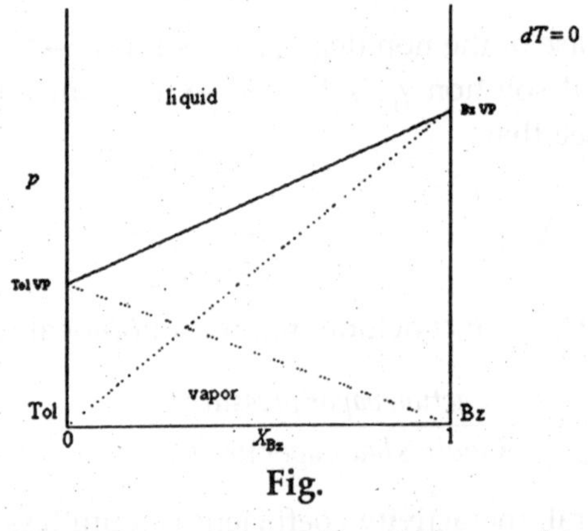

Fig.

On this diagram the dotted line that runs from "Tol VP" to lower right Bz corner is the Raoult's law vapour pressure of toluene. The dotted line that runs from the lower left Tol corner to "Bz VP" is the Raoult's law vapour pressure of benzene. The solid line from "Tol VP" to "Bz VP" is the total vapour pressure of the solution which is just the sum of the two Raoult's law vapour pressures (the sum of two straight lines is a straight line).

There is no indication on the above diagram what the composition of the vapour is at each pressure. We know that at any given pressure the vapour will be richer in the more volatile component, in this case richer in benzene. In our initial discussion of Raoult's law we worked a problem in which the composition of a benzene-toluene solution was 0.400 mole fraction benzene, but the vapour in equilibrium with the liquid was 0.649 mole fraction benzene. This means that we must draw another line to indicate the composition of the vapour at each pressure.

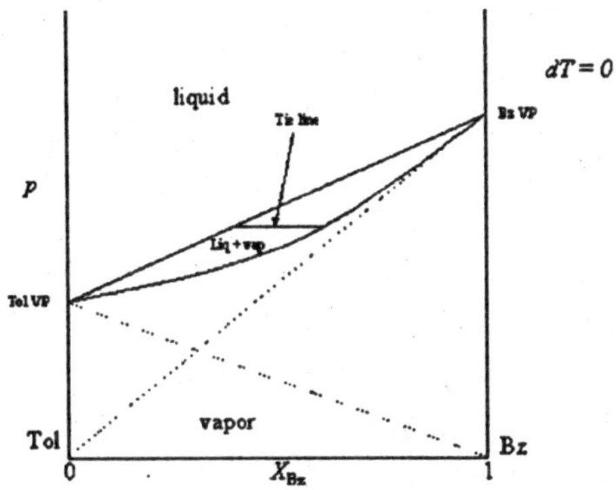

Fig.

In figure we have added the lower curved line from "Tol VP" to "Bz VP" to show the composition of the vapour at each pressure. The region between the curved line and the straight total vapour pressure line is a two phase region. A single point inside this region does not correspond to a state of the system because in the two-phase region we have a vapour phase and a liquid phase in equilibrium with each other. Thus, a point in this region must correspond to two different compositions. The way we find these two different compositions is to draw a "tie line."

Although a point inside the two phase region does not give us the composition of either the liquid or the vapour phase, it does give the overall composition of the whole system. The tie line is a horizontal line through that point which intersects the two boundary lines of the two phase region. The intersection of our horizontal line (the tie line) with the boundary line on the liquid side of the region tells us the composition of the liquid and the intersection of the tie line with the boundary curve on the vapour side tells us the composition of the vapour phase.

Now we have to deal with nonideal solutions. We will distinguish two cases: positive deviations from Raoult's law and negative deviations from Raoult's law. Nonideal solutions which have positive deviations from Raoult's law have vapour pressures which are higher than Raoult's law would predict. This is because the attractive forces between the unlike molecules are weaker than the attractive forces between the like molecules.

Nonideal solutions which have negative deviations from Raoult's law have vapour pressures which are lower than Raoult's law would predict. This is because the attractive forces between the unlike molecules are stronger than the attractive forces between like molecules. A vapour pressure diagram for a system with positive deviations from Raoult's law might look qualitatively like the following:

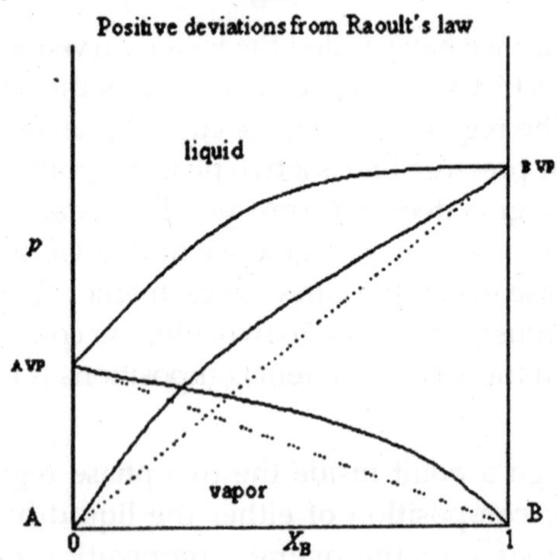

Fig.

This vapour pressure diagram shows qualitatively how the vapour pressures vary with composition for positive deviations from Raoult's law. We have not tried to draw the vapour composition curve on the diagram. We have tried to show that each component obeys Raoult's law in the limit as its mole fraction goes to one. Also, the vapour pressures of the individual components becomes linear in mole fraction as that mole fraction approaches zero. This latter phenomenon is known as "Henry's law" and will discussed further below.

Negative deviations from Raoult's law might look qualitatively like the following diagram.

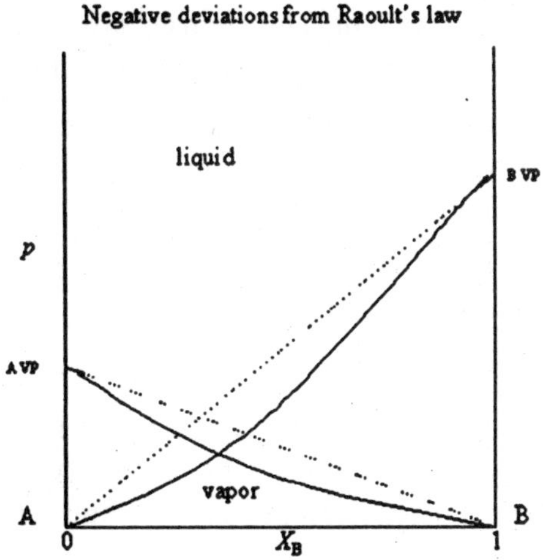

Fig.

Once again we have tried to show that each component obeys Raoult's law as its mole fraction approaches one and becomes linear in its mole fraction as that mole fraction approaches zero. We have not shown the total vapour curve or the vapour composition curve. The total vapour pressure curve would just be the sum of the two individual vapour pressures. Since the solution does not obey Raoult's law it would require experimental data or a more sophisticated theory to construct the vapour composition curve.

HENRY'S LAW

We alluded above to the fact that as $X_i \rightarrow 1$ for each component in a binary mixture the vapour pressure of that component obeys Raoult's law. That is,

$$p_i = X_i p_i^* \tag{1}$$

even for nonideal solutions is the limit as $X_i \to 1$.

However, at the other end of the scale, for small mole fractions, p_i becomes proportional to X_i, but the proportionality constant is not p_i^*. That is, for small mole fractions

$$p_i = K_i X_i^* \tag{2}$$

Equation 2 is called Henry's law and the constant, K_i, is the Henry's law constant. The Henry's law constant depends on the nature of both the solvent and the solute.

It can be shown thermodynamically that in a concentration region where one component obeys Raoult's law the other will necessarily obey Henry's law.

Henry's law is most often used to describe the solubility of gases in liquids.

Example Calculation

Let's calculate the solubility of N_2 gas in water at 25° C and 4.0 atm pressure. From a table of Henry's law constants we find the constant for nitrogen gas in water is 6.51×10^7 Torr. From equation 2 we get

$$X_{N_2} = \frac{P}{K}$$

$$= \frac{4.0 \times 760}{6.51 \times 10^7}$$

$$= 4.7 \times 10^{-5} \tag{3a, b, c}$$

This amounts to about 0.0026 mol of N_2 per liter of water or 0.073 g/L.

One can use Henry's law to calculate the amount of oxygen gas dissolved in water which is exposed to the atmosphere. The Henry's law constant for O_2 gas in water at 25°C is 3.3×10^7. Since air is about 21 mole per cent oxygen, at an air pressure of one atmosphere the partial pressure of O_2

would be about 0.21 atm. It is easy to use Henry's law to calculate the oxygen content of water in contact with air at one atmosphere pressure,

$$X_{O_i} = \frac{P}{K}$$

$$= \frac{0.21 \times 760}{3.30 \times 10^7}$$

$$= 4.8 \times 10^{-6} \qquad \text{(4a, b, c)}$$

This corresponds to about 0.00027 moles of O_2 per liter of water, or 0.0086 g/L.

This does not seem like very much oxygen dissolved in water, but we note two things: Thermodynamically, the chemical potential of the dissolved O_2 is the same as the chemical potential of the O_2 in the gas in equilibrium with the water solution. Second, this seems to be enough to keep fish alive because fish use their gills to extract the oxygen and use it to metabolize food and produce energy.

(There are other issues involved in keeping fish alive in water. At equilibrium there is enough oxygen to keep fish alive, but as the fish remove the oxygen with their gills the concentration of oxygen decreases. The dissolution and diffusion of O_2 through water are relatively slow processes. Unless a way is found to replenish the oxygen the fish will die. In large bodies of water this is no problem, but in small fish tanks it can be fatal to the fish. This is why most fish tanks have a "bubbler" in them. The bubbler increases the exposure of the water to the oxygen in air and gently stirs the water to hasten the saturation of water with oxygen.)

(One other comment about the solubility of gases in water. A condition known to deep sea divers as the "bends" is caused by the solubility of nitrogen in water. We know from Henry's law that the solubility increases with pressure. At the high pressures produced by deep sea diving [and some scuba diving] nitrogen gas dissolves in the blood. If the diver rises from the depths very slowly, over a period of many hours, the blood will gradually release the nitrogen and remain at equilibrium with the surrounding gas pressure. However, if the diver rises too fast the nitrogen will come out of solution as the pressure is released and form bubbles of nitrogen gas in the blood and other body fluids. The bubbles accumulate at the joints, producing great pain, and causing the diver to "double over in pain," Thus the condition is referred to as "the bends." The bends can even lead to death unless a way is found to redissolve the nitrogen gas. Commercial and military diving support ships carry a decompression chamber so that a diver can be brought up quickly and then recompressed in the chamber. [See the James Bond movie, *Licensed to Kill*.] In the decompression chamber the pressure can be released gradually allowing the nitrogen to escape the blood stream without forming the bubbles and making it unnecessary to remain at the diving site to bring the diver up slowly.)

BOILING DIAGRAMS

The diagrams shown above, where we are plotting vapour pressures as a function of composition at constant temperature are not the must useful diagrams. Since we live in a more-or-less constant pressure environment it is more useful and interesting to plot the properties of the solution as temperature versus composition.

Think of the vapour pressure diagrams we showed above as slices at constant temperature through a three-dimensional diagram showing the properties of the mixture as a function of temperature, pressure, and composition. The lines in the above vapour pressure diagrams will appear on this three-dimensional diagram as surfaces, and the points, such as boiling points will be curved lines. (Because the boiling point changes with changing pressure.)

Now, instead of taking slices at constant temperature, let's take slices through our three-dimensional diagram at constant pressure. If we take our slice at a pressure of one atmosphere the new diagram, called a boiling diagram, would look something like,

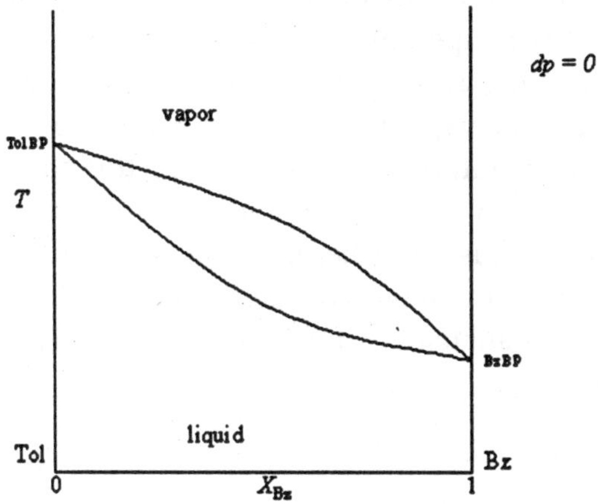

This diagram is called a "boiling diagram." Notice that at the edges we have points representing the boiling points of the pure components. The lower curve gives the boiling point of the liquid mixture as a function of composition.

Suppose we pick a point on the lower curve at, say, a mole fraction of benzene of about 0.33.

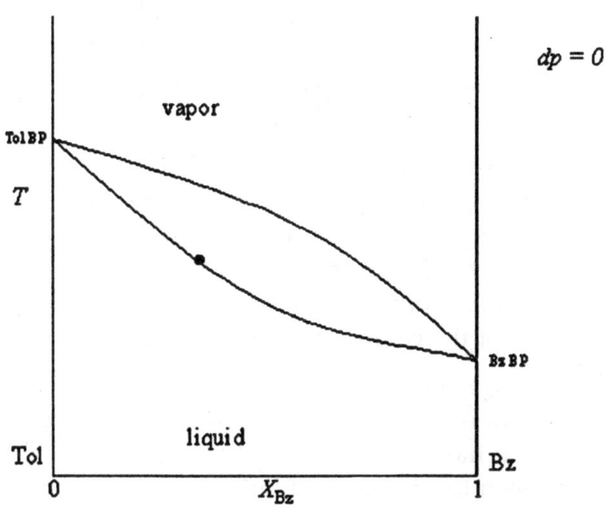

This point gives the boiling point of a mixture which is 0.33 mole percent benzene. We can ask, what is the composition of the vapour which is in equilibrium with the liquid at this temperature? Since we are asking a question about something at this temperature, the temperature of the point, clearly the vapour composition must lie on a horizontal (constant temperature) line going through this point. We know that at any given temperature and pressure the composition of the vapour must be richer in the more volatile component and benzene is the more volatile component. Thus we must draw our constant temperature line from the point proceeding to the right.

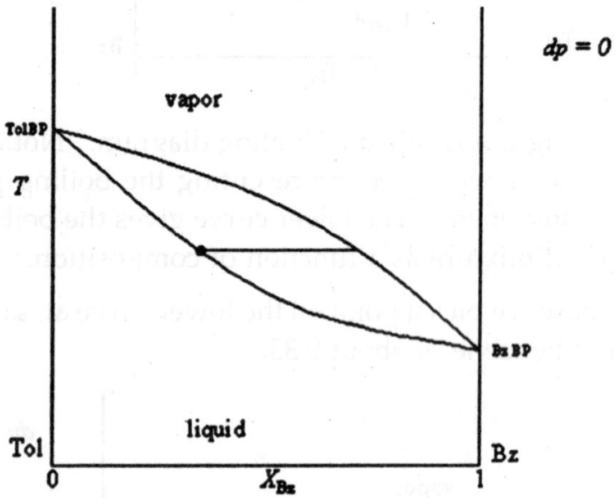

We call this line a tie line and it intersects the upper curve at the composition of the vapour in equilibrium with the liquid at the same temperature. The region between the two curves is a two-phase region. In this region two phases, liquid and gas, are in equilibrium with each other and we must be able to keep track of the compositions of both phases.

Above the upper curve the sample has fully evaporated and the system is entirely a gas. Below the lowest curve none of the sample has evaporated and the system is entirely a liquid. Between the two curved lines the system is composed of two phases in equilibrium and we must draw the tie line to keep track of the composition of both phases.

If we heat a liquid mixture at a composition of about 0.33 mole fraction benzene the liquid will have the composition shown at point "a" when the liquid first begins to boil.

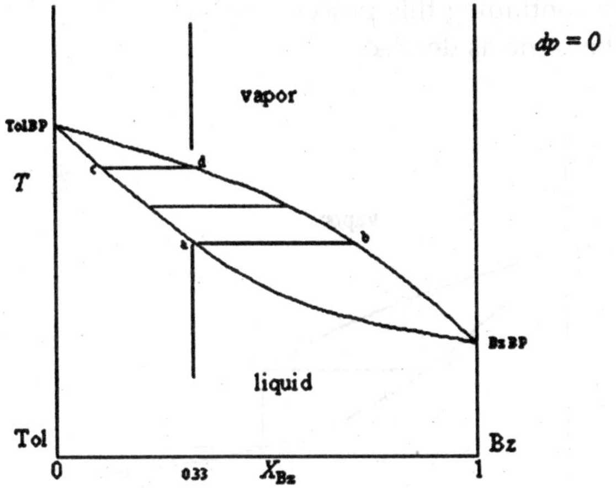

The first vapour to come off will have the composition shown at point "b." As we continue to heat the system the more volatile component will evaporate preferentially and the liquid phase will become richer in the less volatile component (toluene, here). As we continue to heat the mixture the toluene will gradually catch up to the benzene in the vapour phase. The very last bit of liquid to evaporate will have the composition shown at point "c," and the vapour will have the original composition that we started with, "d" (or "a"). Notice that as we heat the system the tie line "tracks" with the lower and upper curves to give the composition of liquid and vapour in equilibrium.

FRACTIONAL DISTILLATION

Boiling diagrams provide an explanation for fractional distillation. Suppose we start with a liquid mixture with 0.33 mole fraction benzene.

As we heat the liquid it will begin to boil when the temperature reaches the temperature of point "a." The first vapour to come off has the composition shown at point "b." Capture the vapour, condense it, and heat it up. The new liquid will boil at point "c" giving a vapour with composition at point "d." Capture this vapour, cool it, and boil it at point, "e," and so on. by continuing this process the vapour can be made as pure in benzene as desired.

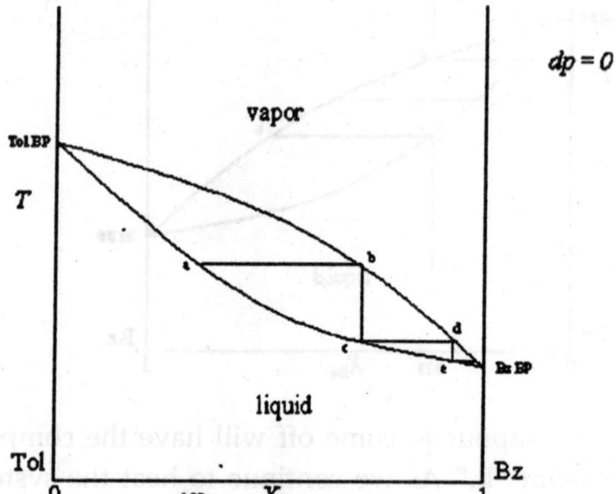

This is what fractional distillation does. There are more modern ways to separate mixtures, for example, chromatography, but fractional distillation still has industrial and some laboratory importance.

BOILING DIAGRAMS FOR NONIDEAL SOLUTIONS

We saw above that a solution with positive deviations from Raoult's law has a higher vapour pressure than would be given by Raoult's law. That means that the solution would boil at a lower temperature than would be expected from Raoult's law. A boiling diagram for such a substance might look like the following:

Point "a" denotes a constant boiling mixture called an "azeotrope." In this case it is a "low boiling azeotrope." We have drawn in some "zig-zag" lines to show why fractional distillation will not give complete separation of the mixture into its components.

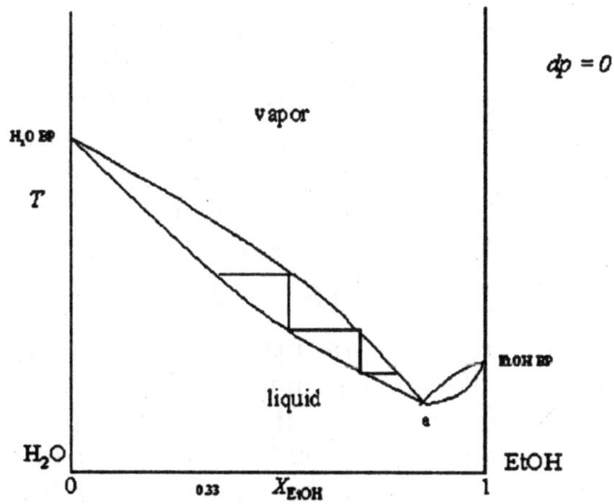

In a mixture with negative deviations from Raoult's law the vapour pressure is lower than would be expected from Raoult's law which will produce boiling points higher than expected and a high boiling azeotrope.

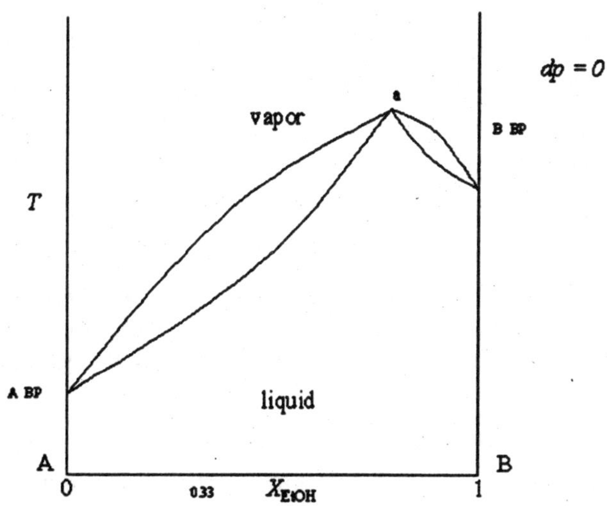

On figure the point "a" indicates the high boiling azeotrope.

It turns out that the composition of the azeotrope is a function of the pressure. One can obtain a solution with very accurately known composition by letting it boil until the high boiling azeotrope is reached. For example, there are tables in the literature that give the composition of the high boiling HCl-H_2O azeotrope as a function of pressure. By measuring the local atmospheric pressure and going to the table you can obtain the composition of the high boiling azeotrope to an accuracy sufficient for acid-base analytical work.

COLLIGATIVE PROPERTIES

Colligative properties are properties of a solution that depend mainly on the relative numbers of particles of solvent and solute molecules and not on the detailed properties of the molecules themselves. You could almost refer to these as statistical properties because they can be understood solely on the basis of counting the relative numbers of particles in a solution. We will derive equations for the colligative properties of ideal solutions. The equations we derive will be valid for ideal solutions and for real solutions in the limit of small concentrations. Nonideal solutions require that corrections be made to these ideal equations because in nonideal solutions the details of intermolecular interactions become important.

We will discuss four colligative properties:

1. Vapour pressure depression
2. Boiling point elevation
3. Melting point depression
4. Osmotic pressure

In all of the following discussion we will denote the solvent as component #1 and the solute as component #2. For example, the mole fraction of component 1 will be symbolized by X_1, and so on.

Vapour Pressure Depression

Vapour pressure depression is the simplest of the colligative properties and the easiest to understand on the basis of a physical model. A given solvent has a vapour pressure which we usually denote by p_1^*. That means that at equilibrium the gas phase above the solvent has a solvent partial pressure of p_1^*. When you add a solute to the solvent to make a solution the partial pressure of the solvent in the gas phase decreases. We can show that this is true using Raoult's law. In this case we have X_1 close to 1 and X_2 small. Then Raoult's law applied to the vapour pressure (partial pressure in the gas phase) of the solvent is,

$$p_1 = X_1 p_1^* = (1 /0\ X_2) p_1^*, \qquad (1)$$

where we have taken advantage of the fact that in a two-component system, $X_1 + X_2 = 1$. Rearrange equation 1 to show the change in the vapour pressure of component 1,

$$p_1 - p_1^* = - X_2\, p_1^*. \qquad (2)$$

We see that the change in the vapour pressure is negative which means that the vapour pressure decreased and it decreased by an amount that is proportional to the vapour pressure of the pure liquid and proportional to the fraction of solute molecules in the solution (that is, the ratio of the number of solute molecules to the total number of molecules).

Vapour pressure depression is relatively easy to understand on the basis of a physical model. At the surface of a liquid there is a competition between the kinetic energy of the molecules (thermal energy), which is trying to push the molecules off the surface into the gas phase, and the intermolecular forces, which are trying to keep the molecules on the surface. (We will see later that the kinetic energy depends on temperature and that the average kinetic energy is proportional to the Kelvin temperature. However, at any given temperature there is a distribution of kinetic energies with most of the molecules having a kinetic energy near the average, but with some higher and some lower. As you increase the temperature you get more molecules with the higher kinetic energy.)

Some of the molecules on the surface with sufficient kinetic energy to overcome the intermolecular forces will escape into the gas phase and contribute to the pressure of solvent molecules in the gas phase. As this pressure increases, there are more molecules in the gas phase colliding with the surface and sticking. At equilibrium the number of molecules leaving the surface just balances those returning to the surface. The measured pressure at this point is the vapour pressure. The solute molecules decrease the vapour pressure because some of the solvent molecules on the surface have now been replaced by solute molecules. There are fewer solvent molecules on the surface to escape and the vapour pressure goes down. You can calculate how much the pressure goes down by counting how many of the solvent molecules on the surface have been replaced by solute molecules.

We can see graphically how the vapour pressure of the solvent decreases with the addition of solute by looking at the vapour pressure diagram below. In this diagram the vapour pressure of the solute (component 1) is plotted against X_1 at constant temperature using Raoult's law. The circled area represents the region of small X_2 we are talking about, although for an ideal solution the straight line extends to X_1 equals zero. The graph in the circled area would be approximately correct even for nonideal solutions because all solutions become ideal with respect to the solvent in the limit as X_1 goes to unity.

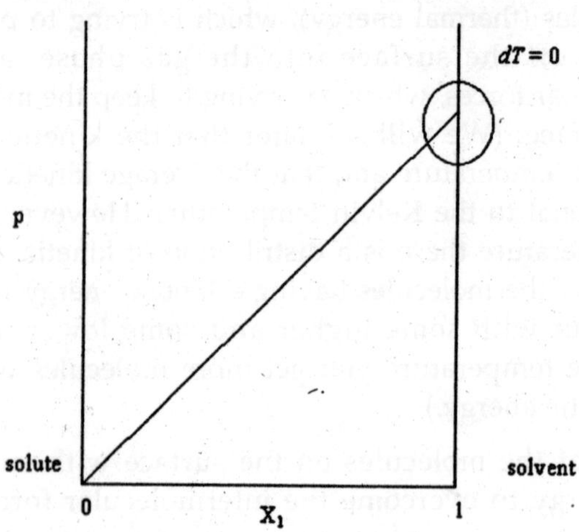

Boiling Point Elevation

It is not hard to understand qualitatively why the boiling point of a solution would be higher than the boiling point of the pure solvent. (We are assuming that the solute is a nonvolatile compound and does not make any significant contribution to the vapour pressure of the solution. If the solute is a volatile compound with a lower boiling point than the solvent the boiling point of the solution will actually go down - by Raoult's law.) We have learned that the vapour pressure of the solvent decreases in a solution. If our solvent was originally at the boiling point ($p_1 = 1$ atm) adding a nonvolatile solute will lower the vapour pressure and the mixture will no longer be boiling. We can bring it back to boiling by increasing the temperature. Thus, the boiling point of the solution is higher that that of the pure solvent.

In the diagram below on the left side we have pure solvent at its boiling point. We know that the chemical potential of the liquid solvent must equal the chemical potential of the solvent in the vapour phase. That is,

$$\mu_{1l} = \mu_{1vap}. \quad (3)$$

Thus, for the vapourization process,

liquid → vapour, > (4)

we have,

$$\Delta\mu = \mu_{1vap} - \mu_{1l} = 0 \quad (5)$$

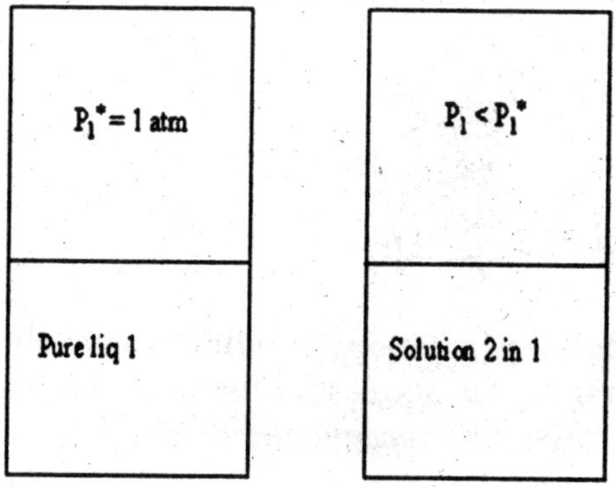

Approximating the vapour as an ideal gas and using the ideal solution equations for the solution we get.

$$\mu_{1vap} = \mu^{\circ}_{1vap} + RT\, In\, p_1 \tag{6}$$

and

$$\mu_{1l} = \mu^{\bullet}_{1l} + RT\, In X_1 = \mu^{\bullet}_{1l} + RT\, In(1 - X_2), \tag{7}$$

where, as usual, $\mu^{\bullet}_{1l}$ is the chemical potential of the pure solvent, is the standard state potential for the gas (vapour), and there is an implied 1 atm dividing p_1 inside the logarithm. Then the difference in chemical potentials can be written,

$$\Delta\mu_1 = \mu^{\circ}_{1vap} + RT\, In\, p_1 - \mu^{\bullet}_{1l} - RT\, In(1 - X_2) \tag{8}$$

We started out, on the left of the diagram, with pure liquid solvent in equilibrium with its vapour at the normal boiling point. Let us now add a small amount of solute, component 2. Add enough to increase the mole fraction of 2 in the solution by dX_2. We insist, however, that the solvent in the solution remain in equilibrium with the solvent vapour. We now ask the question what must the temperature change, dT, be to maintain equilibrium? The answer to this question is contained in the partial derivative,

$$\left(\frac{\partial T}{\partial X_2}\right)_{\Delta\mu_1}. \tag{9}$$

We can figure out how to calculate this derivative by using the cyclic relation

$$\left(\frac{\partial T}{\partial X_2}\right)_{\Delta\mu_1} = -\frac{\left(\dfrac{\partial \Delta\mu_1}{\partial X_2}\right)_T}{\left(\dfrac{\partial \Delta\mu_1}{\partial T}\right)_{X_2}}. \tag{10}$$

The upper derivative on the right is easily obtained from our expression for $\Delta\mu_1$ above and the lower derivative on the left can be obtained by recognizing that,

$$\left(\frac{\partial \mu_1}{\partial T}\right)_p = -\bar{S}_1 \tag{11}$$

or, for the vapourization process,

$$\left(\frac{\partial \Delta\mu_1}{\partial T}\right)_p = -\Delta\bar{S}_1. \tag{12}$$

Then,

$$\left(\frac{\partial T}{\partial X_2}\right)_{\Delta\mu_1} = -\frac{\left(\frac{\partial \Delta\mu_1}{\partial X_2}\right)_T}{\left(\frac{\partial \Delta\mu_1}{\partial T}\right)_{X_2}} = \frac{-RT\frac{1}{1-X_2}(-1)}{-\Delta\bar{S}_{1vap}} \tag{13}$$

We must simplify this expression, but we note first that the above expression requires no specification of the value of $\Delta\mu_1$ other than that it be held constant during the derivative process. However, we know that in order for the solvent to be in equilibrium with its vapour we must have not only $\Delta\mu_1$ = constant, but it must be held at the particular constant value of zero. That is, $\Delta\mu_1 = 0$. In this case,

$$\Delta\mu_1 = \Delta\bar{G}_1 = \Delta\bar{H}_1 - T\Delta\bar{S}_1 = 0, \tag{14}$$

from which we conclude that,

$$T\Delta\bar{S}_{1vap} = \frac{\Delta\bar{H}_{1vap}}{T}. \tag{15}$$

(We have added the subscript "vap" to remind ourselves that we are talking about the vapourization process. With these considerations Equation 13 simplifies to,

$$\left(\frac{\partial T}{\partial X_2}\right)_{\Delta\mu_1} = \frac{RT^2}{\Delta\bar{H}_{1vap}}\frac{1}{1-X_2}. \tag{16}$$

We may need this form of the equation a little later on, but for now we are going to make some approximations. We assume that X_2 is small so that 1 " $X_2 \approx 1$, then,

$$\left(\frac{\partial T}{\partial X_2}\right)_{\Delta\mu_1} = \frac{RT^2}{\Delta\bar{H}_{1vap}}. \tag{17}$$

We can set this up to integrate as follows

$$dT = \frac{RT^2}{\Delta \overline{H}_{1vap}} dX_2. \tag{18}$$

If we are only going to integrate over a small range then our temperature is approximately constant at the boiling point, T_{BP}, and we can write

$$\Delta T = \frac{RT_{BP}^2}{\Delta \overline{H}_{1vap}} \Delta X_2, \tag{19}$$

but $\Delta X_2 = X_2 - 0$, since we started out at zero concentration of solute, so we get

$$\Delta T = \frac{RT_{BP}^2}{\Delta \overline{H}_{1vap}} \Delta X_2. \tag{20}$$

This equation actually solves our problem, but it is not in the form you saw in freshman chemistry. The freshman chemistry form had molality of the solute in place of the mole fraction of solute. To get the form you expect we must relate the mole fraction to the molality,

$$X_2 = \frac{m_2}{\frac{1000}{Mw_1} + m_2}, \tag{21}$$

where m_2 is the molality of solute. That is, m_2 is the number of moles in 1000 g of solvent. (Also, Mw_1 is the formula weight of the solvent.) Since we are assuming a dilute solution the number of moles of solute is much smaller than the number of moles of solvent so we can drop the m_2 in the denominator to get

$$X_2 \approx \frac{m_2}{\frac{1000}{Mw_1}} = \frac{Mw_1 m_2}{1000}. \tag{22}$$

Plugging this expression for mole fraction into the formula we have above for the change in the boiling point we get

$$\Delta T = \frac{RT_{BP}^2}{\Delta \overline{H}_{vap}} \frac{m_2}{\frac{1000}{Mw_1}} = \left[\frac{RT_{BP}^2}{\Delta \overline{H}_{vap}} \frac{Mw_1}{1000} \right] m_2 \tag{23}$$

The expression in square brackets is the boiling point elevation constant (or, if you like big words, the ebullioscopic constant).

If we are dealing with more concentrated solutions we can go back and integrate one of the earlier equations, like

$$dT = \frac{RT^2}{\Delta \overline{H}_{vap}} \frac{1}{1 - X_2} dX_2, \tag{24}$$

or

$$\frac{dT}{T_2} = \frac{R}{\Delta \overline{H}_{vap}} \frac{dX_2}{1 - X_2}. \tag{25}$$

The latter can be integrated from the boiling point of the pure liquid, T_{BP} at $X_2 = 0$ to some new boiling point, T, at concentration X_2. Integration yields,

$$-\frac{1}{T} + \frac{1}{T_{BP}} = -\frac{R}{\Delta \overline{H}_{vap}} In \frac{1}{1 - X_2}. \tag{26}$$

3. Freezing point depression

Freezing point depression is done the same way we did boiling point elevation only this time we must keep the solvent in equilibrium with the pure solid rather than with the vapour. That means we consider the fusion process,

solid -liquid (27)

This time the chemical potential the solid, μ_{1s}, which is the same as the chemical potential of the pure solid solvent, μ_{1s}^*, must be the same as the chemical potential of the solvent in the solution, which we have seen before,

$$\Delta \mu_{1l} = \mu_{1l}^{\circ} + RT\, InX_1 = \mu_{1l}^{\circ} + RT\, In(1 - X_2), \tag{28}$$

but this time

$$\Delta \mu_1 = \mu_{1l}^* + RT\, In(1 - X_2) - \mu_{1s}^*, \tag{29}$$

and

$$\Delta\mu_1 = 0 \tag{30}$$

if the solid and solvent are in equilibrium. Under equilibrium conditions it is true, as before, that

$$\Delta\bar{S}_{fus} = \frac{\Delta\bar{H}_{fus}}{T}. \tag{31}$$

We ask the same question as in our discussion of boiling point elevation. If we add a small amount of solute, component 2, to the solvent to produce an increase in X_2 of dX_2, and we insist that the solid stay in equilibrium with the solvent, what must the temperature change be to maintain the equilibrium. This question is answered by the same type derivative we used in our discussion of boiling point elevation, namely,

$$\left(\frac{\partial T}{\partial X_2}\right)_{\Delta\mu_1} = -\frac{\left(\dfrac{\partial \Delta\mu_1}{\partial X_2}\right)_T}{\left(\dfrac{\partial \Delta\mu_1}{\partial T}\right)_{X_2}} = \frac{RT\dfrac{1}{1-X_2}(-1)}{-\Delta\bar{S}1_{fus}}. \tag{32}$$

This reduces to

$$\left(\frac{\partial T}{\partial X_2}\right)_{\Delta\mu_1} = \frac{RT}{-\Delta\bar{\bar{H}}1_{fus}}\frac{1}{1-X_2}. \tag{33}$$

Notice that we have again used the fact that $\Delta\mu_1$ is not only constant, but constant at the value zero in order to be able to write

$$\Delta\bar{S}1_{fus} = \frac{\Delta\bar{\bar{H}}1_{fus}}{T}. \tag{34}$$

Notice also that this time the derivative in Equation 33 is negative. (You might want to go back and see why we got a negative sign here when we had a positive sign in boiling point elevation.) Under the circumstance where the solution is very dilute, $X_2 << 1$, this reduces to,

$$\left(\frac{\partial T}{\partial X_2}\right)_{\Delta\mu 1} = -\frac{RT_{FP}^2}{\Delta\bar{\bar{H}}1_{fus}}, \tag{35}$$

and we can integrate through a short concentration range as before to get

$$\Delta T = -\frac{RT_{FP}^2}{\Delta \overline{H}_{1fus}} X_2. \tag{36}$$

Again, this is not in the same form you saw in freshman chemistry. To get there we have to convert from mole fraction to molality as before. We get,

$$\Delta T = -\frac{RT_{FP}^2}{\Delta \overline{H}_{1fus}} \frac{m_2}{\frac{1000}{Mw_1}} = -\left[\frac{RT_{FP}^2}{\Delta \overline{H}_{1fus}} - \frac{Mw_1}{1000}\right] m_2. \tag{37}$$

The part in brackets is the freezing point depression constant or the "cryoscopic" constant.

If we want a larger concentration range we have to go back and integrate

$$\left(\frac{\partial T}{\partial X_2}\right)_{\Delta\mu} = -\frac{RT}{-\Delta \overline{H}_{1fus}} \frac{1}{1 - X_2}, \tag{38}$$

to get

$$-\frac{1}{T} + \frac{1}{T_{FP}} = +\frac{RT}{\Delta \overline{H}_{1fus}} IN \frac{1 - X_2}{1}. \tag{39}$$

This expression will be of use later when we talk about solid-liquid phase diagrams.

Osmotic Pressure

The treatment of osmotic pressure is not identical to the treatments of boiling point elevation and freezing point depression we have just given. A cartoon of an osmotic pressure apparatus is shown below. A U-tube has a "semipermeable membrane inserted in the centre of the cross piece as shown. (A semipermeable membrane will allow solvent to flow through it, but not solute.) The left arm of the U-tube is filled with pure solvent and the right arm contains solution. The chemical potential of the solvent is higher in the pure solvent than in the solution so that solvent will flow through the semipermeable membrane from the pure solvent side into the solution side unless it is restrained. (Another way to look at this is that the solution "would like to become more dilute" and it will do so if it can. Given the chance, material will flow from a region of higher chemical potential to a region of lower chemical potential.) To prevent the solvent from flowing we apply an external pressure, Π, to the right side. The pressure, Π, which exactly stops the flow is called the osmotic pressure.

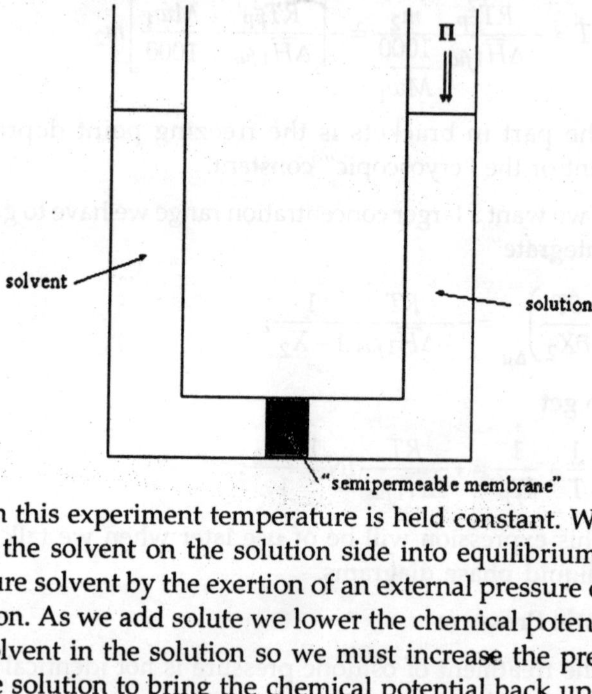

In this experiment temperature is held constant. We will bring the solvent on the solution side into equilibrium with the pure solvent by the exertion of an external pressure on the solution. As we add solute we lower the chemical potential of the solvent in the solution so we must increase the pressure on the solution to bring the chemical potential back up. That is, as we make the solution we need to increase the external pressure to keep μ_{1l} constant. Recall that

$$\mu_{1l} = \mu_{1l}^{*} + RT\,In X_1 = \mu_{1l}^{*} + RT\,In(1 - X_2). \tag{40}$$

The question we ask is: if we increase the concentration of solute by an amount dX_2, how much does the external pressure have to change to keep the solution in equilibrium with the pure solvent? The derivative we want is,

$$\left(\frac{\partial T}{\partial X_2}\right)_{\mu_1} = -\frac{\left(\frac{\partial \mu_1}{\partial X_2}\right)_{p,T}}{\left(\frac{\partial \Delta\mu_1}{\partial T}\right)_{X_2,T}} = \frac{RT\frac{1}{1-X_2}(-1)}{\overline{V}_1}, \tag{41}$$

where we have used the fact that

$$\left(\frac{\partial \mu_1}{\partial p}\right)_T = \left(\frac{\partial \overline{G}_1}{\partial p}\right)_T = \overline{V}_1 \tag{42}$$

Then Equation 41 simplifies to,

$$\left(\frac{\partial p}{\partial X_2}\right)_{\mu_1} = \frac{RT}{\overline{V}_1}\frac{1}{1-X_2} \approx \frac{RT}{\overline{V}_1}. \tag{43}$$

Prepare to integrate this equation,

$$dp = \frac{RT}{\overline{V}_1}dX_2, \tag{44}$$

which integrates to (calling the applied external pressure, or the osmotic pressure, Π),

$$\Pi - 0 = \frac{RT}{\overline{V}_1}(X_2 - 0), \tag{45}$$

or, when the solution is dilute,

$$\Pi = \frac{RT}{\overline{V}_1}X_2 = \frac{RT}{\overline{V}_1}\frac{n_2}{n_1+n_2} \approx \frac{RT}{n_1\overline{V}_1}n_2. \tag{46}$$

But so,

$$\Pi = RT\frac{n_2}{V}. \tag{47}$$

This looks exactly like the ideal gas equation of state, except the pressure is the osmotic pressure, Π, and the system is a solution, not a gas.

This derivation of osmotic pressure assumed the solution is ideal. For ionic solutions, which are not ideal, you can get an approximate osmotic pressure by taking account of the dissociation of the compound into ions in solution. If the compound gives c ions per compound unit ("molecule") the osmotic pressure for dilute ionic solutions can be written,

$$\Pi \approx cRT\frac{n_2}{V}. \tag{48}$$

For more accurate work you can write an expansion similar to the virial expansion,

$$\frac{\Pi V}{n_2 RT} = 1 + B\frac{n_2}{V} + C\left(\frac{n_2}{V}\right) + \ldots, \tag{49}$$

or

$$\frac{\Pi V}{cn_2 RT} = 1 + B\frac{n_2}{V} + C\left(\frac{n_2}{V}\right)^2 + \ldots, \tag{50}$$

for ionic solutions.

Osmotic pressure is enormously important for life. For example, cell wall membranes, in many cases, are semipermeable membranes which will pass molecules of water but not solute molecules or ions. As a consequence the concentrations of material in the fluids inside and outside the cell must be carefully balanced so that they generate the same osmotic pressure. Otherwise, the difference in solute concentrations would generate a pressure differential across the cell wall and either collapse the cell or burst it. Fluids used in intravenous feeding, drug delivery, or just plain fluid replacement are called isotonic solutions because they are designed to have the same osmotic pressure as normal blood cells. The typical isotonic solution is 8.9 g NaCl per liter of water. If a blood cell is placed in a solution with a lower concentration than this (a hypotonic solution) fluid moves from the solution into the cell and cell bursts (called hemolysis). If a blood cell is placed in a solution with a higher concentration (a hypertonic solution) fluid moves from the cell to the surrounding solution and the cell shrinks and dies (called crenation).

It is instructive to calculate the osmotic pressure of an isotonic solution (8.9g NaCl per liter of water). This is the pressure that would be generated inside a blood cell that was placed in pure water. Do you suppose that the cell wall could withstand this pressure?

GIBBS PHASE RULE

Let's try to motivate the Gibbs phase rule. Consider a typical one"component phase diagram.

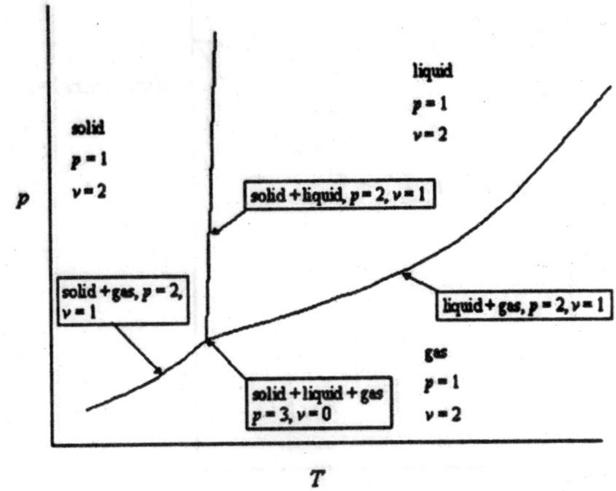

In the one"phase regions one can vary either the temperature, or the pressure, or both (within limits) without crossing a phase line. We say that in these regions there is a variance of 2. We have indicated in the solid, liquid, and gas regions that there is one phase and the variance is two. Along a phase line we have two phases in equilibrium with each other, so on a phase line the number of phases is 2. However, if we want to stay on a phase line, we can't change the temperature and pressure arbitrarily. If we change the temperature " and keep two phases " then the pressure must change also to keep us on the phase line. Otherwise we go off the line and we no longer have two phases in equilibrium. So on a phase line the number of phases is 2, but the variance is 1. At the triple point there are three phases in equilibrium, but there is only one point on the diagram where we can have three phases in equilibrium with each other. Therefore, at the triple point the variance is zero. Notice that the variance seems to be related to the number of phases, such as,

$v = 3 - p$, or (1)

$v = 2 + 1 - p.$ (2)

(You will see why we wrote it this way in a minute.)

Now let's look at another example. Consider the boiling diagram of a two"component ideal liquid"liquid solution.

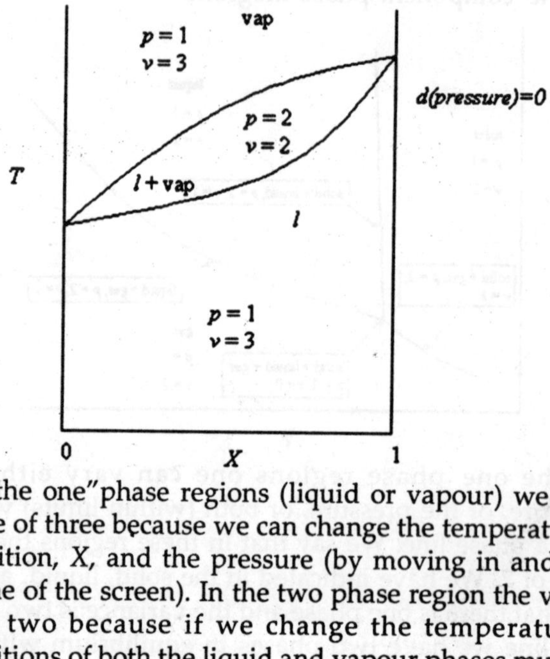

In the one"phase regions (liquid or vapour) we have a variance of three because we can change the temperature, the composition, X, and the pressure (by moving in and out of the plane of the screen). In the two phase region the variance is only two because if we change the temperature the compositions of both the liquid and vapour phases must track as indicated by the tie"lines connecting the liquid and vapour composition lines. Here it looks like the variance satisfies,

$v = 4 - p$, or (3)

$v = 2 + 2 - p.$ (4)

Notice that equations 2 and 4 are the same except for the second term. In Equation 2 there is one component and the second term is 1. In Equation 4 there are two components and the second term in Equation 4 is a 2. It looks like the second term may indicate the number of components in the system. We might guess that the general form for Equations 2 and 4 might be,

$$v = 2 + c - p.$$

We shall see below whether or not our guess is correct.

Before we try to derive a general form for the variance of a system in equilibrium we need to take a look as some mathematical principles.

Take a system where we have three variables, x, y, and z. Let's look at five cases:

Case 1. There is no equation connecting the variables. Then we can pick the values of all three variables to be anything we wish, independently of each other. There are no restrictions on the values of the variables and the variance is 3.

Case 2. There is one equation connecting the variables,

$$x + 2y + 3z = 10 \tag{5}$$

In this case we can select the values any two of the variables, but when we do, the value of the third is fixed by the equation. The number of equations is one and the variance is 2.

Case 3. There are two equations connecting the variables,

$$x - 2y + 3z = 10$$

$$x + y + z = 6. \qquad \text{(6a, b)}$$

In this case we can select the value of any one variable at random, but then the values of the other two are fixed by the two equations. (Two equations in two unknowns is a soluble problem.) The number of equations is two and the variance is 1.

Case 4. There are three equations connecting the variables,

$$x - 2y + 3z = 10$$

$$x + y + z = 6$$

$$x + 3y - z = 8. \qquad \text{(7a, b, c)}$$

In this case we can not select any of the variables arbitrarily. The values of the variables are fixed because a system of three unknowns and three (linearly independent) equations has a unique solution.

$x = 3$

$y = 2$

$z = 1$ (8a, b, c)

The number of equations is three and the variance is zero.

Case 5. There are four linearly independent equations connecting the variables.

$x - 2y + 3z = 10$

$x + y + z = 6$

$x + 3y - z = 8$

$x + 2y + 2z = 7.$ (9,a, b, c, d)

This system does not have a solution. There are no values of the three variables which will satisfy all four equations. We conclude that we cannot have a variance less than zero.

The principle we extract here is that

variance = number of variables – number of equations ≥ 0. (10)

Let us now apply this principle to phase equilibrium. Consider a system which is composed of *c* components. That is,

c = number of components. (11)

(We define the number of components as the minimum number of independent substances (compounds and/or elements) required to form the system. If you can make one substance out of the others you can't count it! For example, a system containing CO_2, CaO, and $CaCO_3$ has two components because you can form the $CaCO_3$ by reacting the other two. There is, however, a special case. If our system contains exactly the same number of moles of CaO and CO_2 it is a one component system because it can be made starting with just pure $CaCO_3$. That is, you can heat the $CaCO_3$ to generate the other two in equal molar amounts.)

Number the components 1, 2, 3, . . . *c*.

Let

p = number of phases. (12)

(We will have to be careful to distinguish between the number of phases and the pressure since we are using the same symbol.)

Enumerate the phases by lower case Greek letters, α, β, γ, δ, . . . ζ. Let's now count the number of variables and the number of equations. We will get the variance from

v = variables – equations. (13)

Our variables are T, p, (pressure) and the mole fractions of the components in each phase.

T,	p,		
$X_{1\alpha}$	$X_{1\beta}$	$X_{1\gamma}$	...
$X_{2\alpha}$	$X_{2\beta}$	$X_{2\gamma}$	...
-	-	-	-
-	-	-	-
-	-		

(14)

Notice that there are p columns and c rows in the array of mole fraction variables. That is, there are cp mole fraction variables. This gives a total of 2 + cp variables.

Now let's count the equations. First of all the mole fractions in each phase must sum to unity. That is

$$X_{1\alpha} + X_{2\alpha} + X_{2\alpha} + \dots\dots\dots = 1$$
$$X_{1\beta} + X_{2\beta} + X_{2\beta} + \dots\dots\dots = 1 \quad (15)$$

Since there is one such equation for each phase this gives p equations. Next, the chemical potential of each component must be the same in every phase. (In the following we designate ζ as the last phase, that is the phases are denoted α, $\beta, \gamma, \ldots, \zeta$.)

$$\begin{array}{cccc} \mu_{1\alpha} = \mu_{1\beta} & \mu_{2\alpha} = \mu_{2\beta} & \cdots & \mu_{c\alpha} = \mu_{c\beta} \\ \mu_{1\beta} = \mu_{1\gamma} & \mu_{2\beta} = \mu_{2\gamma} & \cdots & \mu_{c\beta} = \mu_{c\gamma} \\ - & - & - & - \\ - & - & - & - \\ - & - & - & - \\ \mu_{1(\zeta-1)} = \mu_{1\zeta} & \mu_{2(\zeta-1)} = \mu_{2\zeta} & & c(\zeta-1) = \mu_{c\zeta} \end{array} \tag{16}$$

(Note that there is no row of the form $\mu_{1\zeta} = \mu_{1\alpha}$, etc. Equations of this form are not linearly independent because they can be deduced from the other $p - 1$ equations.) So for this array of chemical potential equations there are c columns and p " 1 rows, which yields $c(p$ " 1) equations. The total number of equations is then $p + c(p$ " 1). If, as we have argued above, the variance is given by

$$v = \text{variables} - \text{equations}, \tag{13}$$

we get,

$$v = 2 + cp - p - c(p - 1) \tag{16}$$

$$v = 2 + c - p. \tag{17}$$

Equation 17 is called the Gibbs phase rule and it is a generally valid thermodynamic equation. That is, it contains no approximations. If you go back and check it against the (motivational) examples we started with you will see that it agrees exactly with our initial observations.

The Gibbs phase rule is an important tool in the analysis of systems with several phases and one or more components. It is used not only in chemistry, but also in geology and geophysics

TWO-COMPONENT PHASE DIAGRAMS

We have previously looked at two-component boiling diagrams. These diagrams are descriptions of the state of the system on a graph of temperature versus composition (at constant pressure). At high temperatures the system is all in the gas (or vapour) phase. At the lower temperatures the system is in the liquid phase. In between these two situations there is a region where there are two phases (vapour and liquid) in equilibrium with each other.

If we continue to cool the system we will eventually reach a temperature where one or both of the pure components will freeze. At temperatures at and below the melting points the phase diagram will look something like the following hypothetical phase diagram involving a substance "A" and a substance "B." In this diagram we are plotting temperature versus the mole fraction of substance B.

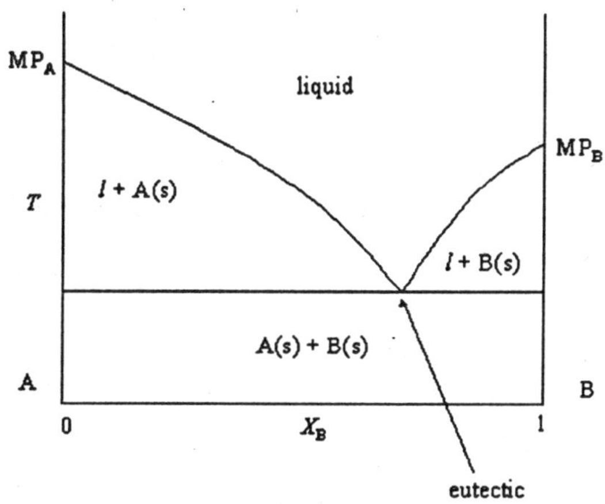

Note that the freezing point of the solution decreases as we move away from either pure A or pure B. This is a phenomenon that we have discussed previously when we talked about freezing point depression in the section on colligative properties.

Let's pick a point in the liquid region on the above diagram and cool the system at constant composition and pressure along the dotted line starting at point "a."

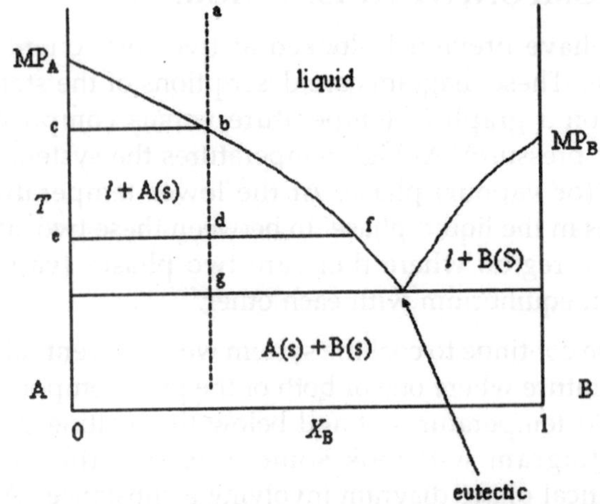

Going from point a to point b nothing much happens. We are just cooling the liquid (or melt). At point b, however, we begin to crystallize out some pure component A. When we cross the curved line at point b we are moving into a two phase region. On this diagram the two phases are pure component A and the liquid mixture. As we already know, in a two phase region we must track two different compositions so we draw a tie-line. In this case the tie-line runs from point b to point c. Where the tie-line intersects the closed curve at b indicates the composition of the liquid phase. In this diagram the other end of the tie-line intersect the edge of the diagram at pure solid A (point c). As we continue to cool the system the tie-lines track the composition of the liquid and solid phases. By the time we get to point d the left end of the tie-line (point e) still tracks the pure solid A. The right end of the tie-line tracks the liquid composition at point f. Notice that the liquid at point f is much richer in component B. This is because we have been removing component A from the solution by crystallizing it out.

When we reach point g the liquid has reached the eutectic composition. At this temperature and liquid composition both pure solid A and pure solid B will crystallize out together. In this phase diagram crystals of the two substances will be interspersed among each other. (The crystals may very well be microscopic so that the mixture looks homogeneous to the naked eye. In this case; the only way to tell that we have a mixture of two different crystals is to examine the solid under a microscope.) If we continue to remove heat from the mixture the system will remain at the eutectic temperature until all of the remaining liquid has solidified.

As the temperature decreases below point g the system crosses over into another two phase region. This time the two phases in equilibrium are pure solid A and pure solid B. Tie-lines in this region (which we have not drawn) run from the pure A side to the pure B side.

The Gibbs phase rule tells us the variance in the various regions. Since this is a two-component system the variance is given by $v = 2 + c - p = 4 - p$. One of the degrees of freedom contributing to the variance is pressure, but since pressure is constant in this diagram we have to regard pressure as being plotted perpendicular to the plane of the screen. It is usual to define the "reduced variance," v', as just the variance in the plane of the diagram. That is, the reduced variance is the phase rule variance minus the pressure variable. The reduced variance is given in a two-component phase diagram by $v' = 3$ " p. In the above diagram the reduced variance is 2 in the liquid region. (You can change temperature and composition arbitrarily over a reasonable range.) In the two-phase regions, like the A(s) + liquid, or B(s) + liquid, regions the reduced variance is 1. In these regions if you change the temperature the composition of the liquid phase must change. (Although the composition of the solid phase will not change in these cases because the solid phase is pure A or pure B.)

SOLID/SOLID SOLUBILITY

In many cases when we cool the melt to its "freezing point" it is not one of the pure components that crystallizes out, but a solid solution. A solid solution has many of the characteristics of a liquid solution except that it is solid. The primary similarity is that in a solid solution the "solute" is distributed homogeneously throughout the "solvent" as atoms or molecules (not as microcrystals). On the diagram below the region labeled α(s) is a solid solution of B in A. That is, component A would be regarded as the solvent and B as the solute. Some people would refer to this phase as "impure A." Likewise, the region labeled β(s) is a solid solution of A in B. It could also be regarded as impure B.

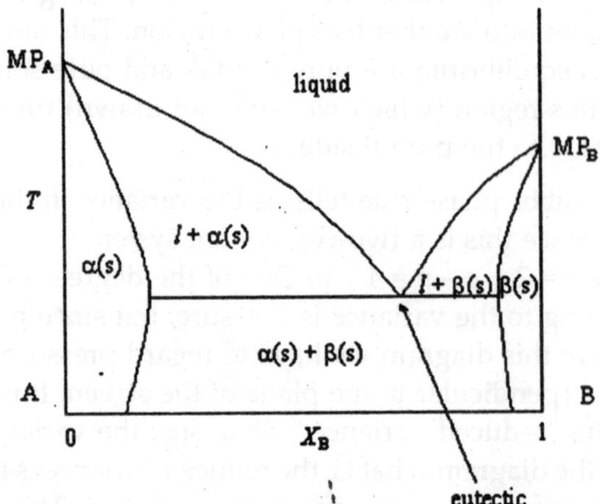

The reduced variance in the solid solution regions labeled α(s) and β(s) is 2 because these regions are one-phase regions. The reduced variance in the other regions is the same as in the phase diagram with no solid-solid solubility. On this diagram it is easier to see why the reduced variance is 1 in the region labeled α(s) + β(s). The composition of the two solid solution phases in equilibrium changes slightly as the temperature is changed.

Silver and copper form solid-solid solutions similar to the above diagram where the solid-solid solubility of one component in the other is not very large.

Diffusion in solids is very slow at room temperature. Diffusion is much faster at temperatures near the melting points of the components. There is a sculpture in a Manhattan park entitled *3000* A. D. *Aluminum/Magnesium* by Terry Fugate-Wilcox. It is a spire of alternating pieces of aluminum and magnesium. Presumably solid-solid diffusion will transform the sculpture into a homogeneous solid-solid solution of aluminum and magnesium by the year 3000.

COMPOUND FORMATION

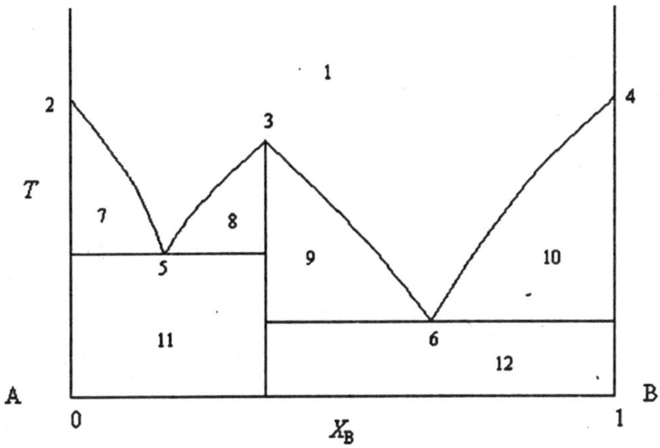

The above phase diagram shows the formation of a compound between A and B. The compound is shown by the vertical line from point 3. You can tell the formula of the compound from its composition, X_B. It looks like the compound has a mole fraction of B of 1/3. That means that in the compound,

$$X_B = \frac{n_B}{n_A + n_B} = \frac{1}{3}.$$

To find the formula of the compound we look for the smallest whole numbers for n_A and n_B which will satisfy this equation. In this case it looks like the equation is solved by setting $n_B = 1$ and $n_A = 2$. Then,

so the formula of the compound is A_2B.

The diagram is labeled as follows:

1 = liquid

2 = MP of A

3 = MP of compound A_2B

4 = MP of B

5 = eutectic of A and A_2B

6 = eutectic of A_2B and B

7 = A(s) + liquid

8 = liquid + A_2B(s)

9 = A_2B(s) + liquid

10 = liquid + B(s)

11 = A(s) + A_2B(s)

12 = A_2B(s) + B(s)

Incongruent Melting Point (melting with decomposition)

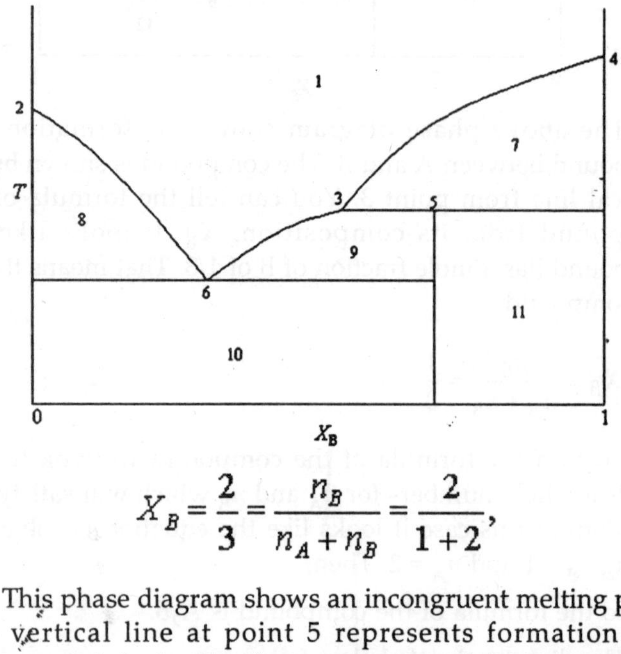

$$X_B = \frac{2}{3} = \frac{n_B}{n_A + n_B} = \frac{2}{1+2},$$

This phase diagram shows an incongruent melting point. The vertical line at point 5 represents formation of a compound. It looks like the composition of the compound is X_B = 2/3, from which we conclude using the previous arguments, so that the compound is AB_2. Point 5 is the melting point of AB_2, but notice that melting AB_2 does not give liquid of the same composition. Rather, melting of AB_2 gives liquid with the composition at point 3 and pure B(s). So the compound, AB_2, melts and decomposes at the same time. An analysis of the points and regions is:

1 = liquid

2 = MP of A

3 = the peritectic point (we don't get a eutectic of AB_2 and B)

4 = MP of B

5 = the incongruent melting point of AB_2

6 = eutectic of A and AB_2

7 = liquid + B(s)

8 = A(s) and liquid

9 = liquid and AB_2(s)

10 = A(s) + AB_2(s)

11 = AB_2(s) + B(s)

OTHER POSSIBILITIES

There are many possible variations and combinations possible with these themes. For example,

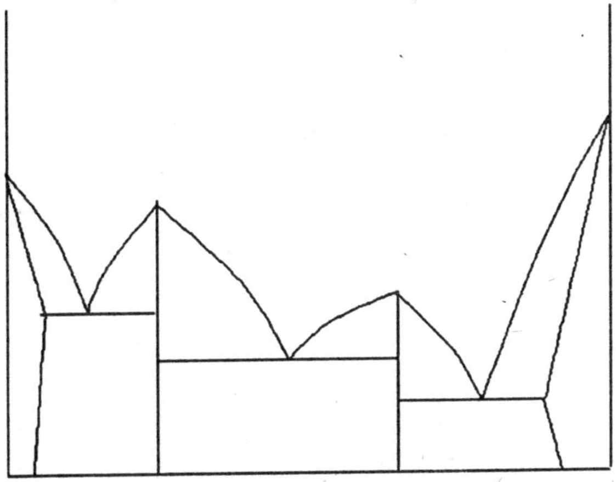

Or even more interesting,

We will leave it to the reader to understand and analyze these two hypothetical phase diagrams.

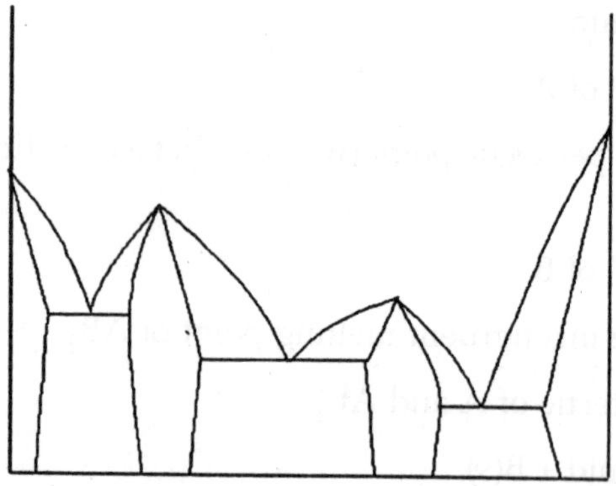

Phase diagrams, some much more complicated than the examples presented here, are very important in areas such as metallurgy, materials science, geology and geophysics, planetary science, and many more. Some of the most important ones involve more than two components. These are beyond the scope of the present discussion.

COOLING CURVES

It is fair to ask what kind of experimental data it takes to determine the phase diagram of a two-component system. One such data set comes from cooling curves. To make a cooling curve you start with liquid at some composition. Then you allow it to cool slowly and measure the temperature as a function of time. (You want it to cool slowly so that the system has time to establish equilibrium at each temperature. This may be hard to do because physical and chemical processes involving solids can be very slow. Nevertheless, you would like to be as close to equilibrium as you can be.)

If you start at the composition of a pure compound then the system cools relatively rapidly until you come to the melting point of the compound. At the melting point the temperature holds steady at the melting temperature until all the liquid has been converted to solid. This is called a "halt." Then the temperature drops rapidly again. The same thing happens at the composition of a eutectic. The temperature drops relatively rapidly until you hit the eutectic melting point and then "halts" until all the liquid has been converted to solid, then the temperature drops rapidly again.

If you start at a composition between a pure compound and the eutectic the temperature drops relatively rapidly until you hit the temperature of one of the solid/liquid phase boundaries. At this point the mixture continues to cool, but much more slowly because the system is gradually precipitating out a pure compound as it cools and the heat of fusion must be dissipated in order for the system to cool.

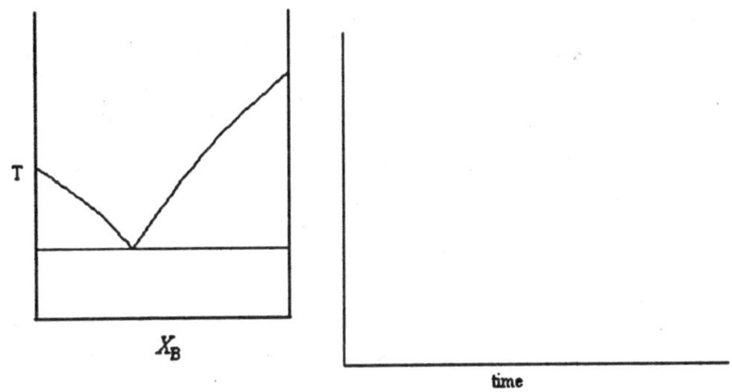

The diagram above shows, on the left, a simple hypothetical phase diagram (substance A with substance B). On the right there is a set of coordinate axes for a plot of temperature versus time (using the same temperature scale). Since this stupid Microsoft Word/Powerpoint combination will not let me place all the cooling curves on one graph I will show you five graphs of cooling curves at different compositions.

The first cooling curve will be for pure A:

As we cool pure liquid A nothing much happens until we reach the melting point of A. At the melting point of A the curve takes a "halt." That is, the temperature holds constant until all of the liquid has been converted into solid. The temperature stays constant for a period of time because our cooling mechanism, whatever it is has to remove the heat of fusion which is given off when the liquid is convert into solid. When all of the liquid has been converted to solid we begin cooling the solid and nothing else of interest happens (unless there is some solid-solid phase transition at a lower temperature which is not shown on our graph.)

Our second cooling curve will be for a mixture with composition between pure A and the eutectic.

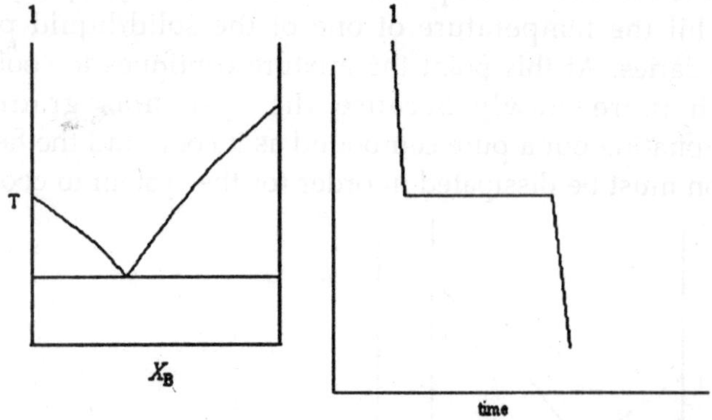

On this cooling curve we have a break (change in slope) at the point where A(s) begins to precipitate out. The system cools slower at the break because A(s) is crysalizing (freezing) out and the heat of fusion being released in this process must be dissipated in addition to the heat removed in simply lowering the temperature.

The next cooling curve will be at the composition of the eutectic. Nothing happens until we reach the melting point of the eutectic, at which point the temperature remains constant until all of the material has solidified.

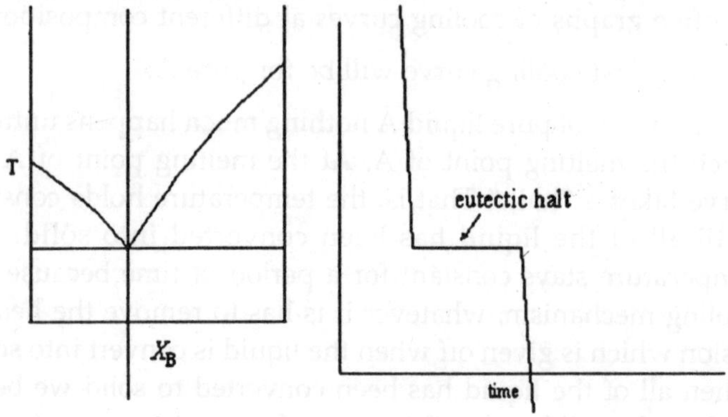

Our fourth cooing curve will be at a composition between the eutectic and pure B. We expect to see a break at the point where the temperature crosses the liquid/solid phase boundary and a halt at the melting point of the eutectic.

And we do.

Cooling curve will be for pure B.

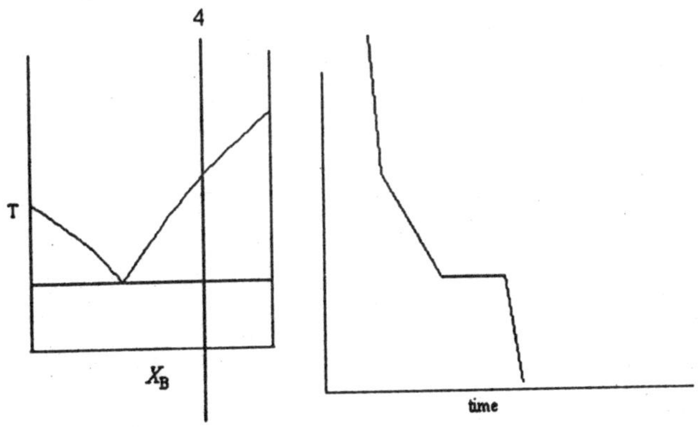

All we get for the cooling curve of pure B is a halt at the melting point of B.

It is not so much of interest to go from the phase diagram to a set of cooling curves (although you should be able to do that). The main interest is to get the phase diagram from a set of cooling curves. Experimental data is often in the form of a set of cooling curves (or even in a specification of the breaks and halts at different compositions). From this data one can construct the phase diagram.

CONCENTRATION CELLS [H]

It is not always necessary to have a net chemical reaction in order to generate a cell voltage. A concentration cell is an example of a cell which will produce a voltage (albeit a small voltage) without a net chemical reaction. Consider the cell,

$$Zn(s) | ZnSO_4(aq, m_L) || ZnSO_4(aq, m_R) | Zn(s).$$

(Note that the "L" and "R" refer to the left- and right-hand sides of the cell respectively.) We analyze this cell as follows:

R (= Reduction) $Zn^{2+}(aq,m_R) + 2\,e'' \rightarrow Zn(s)$

L (= oxidation) $Zn(s) \rightarrow Zn^{2+}(aq,m_L) + 2\,e''$.

The overall reaction is just the sum of these two half-reactions:

$Zn^{2+}(aq,m_R) \rightarrow Zn^{2+}(aq,m_L)$.

(Note that the zinc ion concentration from the **right** hand side of the cell appears to the left of the arrow in the balanced overall reaction and vice versa.)

The E^o for this reaction is given, as usual, by,

$$E^\circ = E^\circ_R - E^\circ_L$$

but in this case the oxidation half reaction is just the reverse of the reduction half reaction so that E^o_R and E^o_L are the same and the overall E^o is zero.

Nevertheless, we can generate a voltage if the concentrations on each side are the different. Using the Nernst equation we get

$$E = E^\circ - \frac{RT}{2\zeta}\,\mathrm{In}\frac{a_L}{a_R}$$

$$= 0 - \frac{RT}{2\zeta}\,\mathrm{In}\frac{\dfrac{m_L^2\gamma^2 \pm L}{m_R^2\gamma^2 \pm L}}{m_R^2\gamma^2 \pm R}$$

This makes sense, because if the concentrations on the two sides are the same then the voltage is zero. However, if the concentration on the right side of the cell (not the right side of the reaction) is larger then the argument of the logarithm is less than one so that its logarithm is negative and E is positive. We know from other considerations that there is a natural drive toward dilution so that this process is spontaneous.

In principle any isothermal change in Gibbs free energy can be set up to generate a voltage. In this case the Gibbs free energy change in going from a more concentrated solution to a more dilute solution is negative. This negative Gibbs free energy change can be used to give a voltage and produce electrical energy.

HOW TO MEASURE E^o AND $\gamma_\pm$ [H]

We will show how to measure E^o and $\tilde{a}_\pm$ by means of an example. Consider the cell

Pt(*s*) | H_2(*g*, *p*) | HCL (*aq*, *m*) | AgCl(*s*) | Ag(*s*)

Analyze it as usual, but now we assume the only emf that we know is that of the standard hydrogen electrode. We want to measure the emf of the AgCl(*s*) | Ag(*s*) half-cell. Call it E^o.

R $AgCl + e^- \rightarrow Ag + Cl^-$ $\qquad E_R{}^o = E^o$

L $\frac{1}{2} H_2 \rightarrow H^+ + e^-$ $\qquad E_L{}^o = 0$

$AgCl + \frac{1}{2} H_2 \rightarrow Ag + HCl(aq, m)$

In the above reactions the E^o is one of the quantities we want to measure. The left-hand electrode is just the standard hydrogen electrode. We use the Nernst equation approximating H_2 as an ideal gas (a pretty good approximation at low pressures).

$$E = E^\circ - \frac{RT}{2\zeta} In \frac{a_{Hcl(aq)}}{a_{Agcl(s)} \left(\frac{pH_2}{1\,\text{atm}} \right)^{\frac{1}{2}}}$$

$$= E_0 - \frac{RT}{\zeta} In \frac{m^2 \gamma_\pm^2}{PH_2}. \qquad \text{(1a, b)}$$

Now hold the pressure of H_2 constant at 1 atm. Rewrite Equation 1b as,

$$E = E^\circ - \frac{RT}{2\zeta} \text{In}\, m^2 \frac{RT}{\zeta} In\gamma H_\pm^2$$

$$= E_0 - \frac{RT}{\zeta} \text{In}\, m - \frac{2RT}{\zeta} In\gamma_\pm. \qquad \text{(2ab)}$$

Rearrange Equation 2b as follows:

$$E^{\circ} = \left(E + \frac{2RT}{\zeta} \ln m \right) + \frac{2RT}{\zeta} \ln \gamma_{\pm} \tag{3}$$

There are two unknowns in this equation E° and $\tilde{a}_{\pm}$, but we prepare our solutions so that we know the molality, m, and we measure E at each value of m. Also, we know that $\tilde{a}_{\pm} = 1$ as $m = 0$.

We find E° by plotting the quantity $E + \frac{2RT}{\zeta} \ln m$ vs m and extrapolate to $m = 0$. This curve intercepts the $m = 0$ axis at E°.

This might be hard to do in practice because it is tricky to work with very small concentrations. Also, note that ln m gets very large and negative as m gets small. so presumably E is getting large and we are subtracting two large numbers to get a small one.

An improvement can be made by using the Debye-Hückel limiting law.

$$\ln_{\pm} = -1.177\sqrt{I}, \tag{4}$$

where $I = m$ for HCl(aq) so that

$$\ln \gamma_{\pm} = -1.177\sqrt{m}, \tag{5}$$

Then

$$E^{\circ} = \left(E + \frac{2RT}{\zeta} \ln m \right) + \frac{2RT}{\zeta} \left(-1.177\sqrt{m} \right)$$

$$= \left(E + \frac{2RT}{\zeta} \ln m \right) - \frac{2RT}{\zeta} 1.177\sqrt{m} \tag{6a, b}$$

Remember that E and m are measured.

Plot $E + \frac{2RT}{\zeta} \ln m - \frac{2RT}{\zeta} 1.177\sqrt{m}$ vs m and extrapolate to $m = 0$. The intercept is still E°, but the extrapolation should be easier to carry out.

Once we know E^o we can obtain $\tilde{a}_{\pm}$ at any m by measuring E at that m. From,

$$E = E^o - \frac{2RT}{\zeta}\ln m - \frac{2RT}{\zeta}\ln\gamma\pm \tag{7}$$

we solve for ln $\tilde{a}_{\pm}$ as,

$$In\gamma_{\pm} = \left(E^o - E\right)\frac{\zeta}{2RT} - Inm \tag{8}$$

We can use these measurements to make tables of $\tilde{a}_{\pm}$ or we can fit the data to a curve.

THE CYCLIC RELATION [H]

The "cyclic relation" (sometimes called Euler's chain relation) is a calculus identity which is very useful in thermodynamics. This relation can appear in several different forms all of which are equivalent. The form that we will find most useful is,

$$\left(\frac{\partial x}{\partial y}\right)_z = -\frac{\left(\frac{\partial z}{\partial y}\right)_x}{\left(\frac{\partial z}{\partial x}\right)_y}$$

We will use this in many different contexts, but one example would be the calculation of the derivative, $\left(\frac{\partial V}{\partial T}\right)_p$ using the van der Waals equation of state. The derivative is difficult to calculate directly because solving for V as a function of p and T requires solving a cubic equation in V. One can obtain this derivative from the cyclic relation by calculating,

It is relatively easy to solve the van der Waals equation of state for p in terms of V and T and to take the appropriate derivatives shown.